CW01084079

Springer Theses

Recognizing Outstanding Ph.D. Research

Aims and Scope

The series "Springer Theses" brings together a selection of the very best Ph.D. theses from around the world and across the physical sciences. Nominated and endorsed by two recognized specialists, each published volume has been selected for its scientific excellence and the high impact of its contents for the pertinent field of research. For greater accessibility to non-specialists, the published versions include an extended introduction, as well as a foreword by the student's supervisor explaining the special relevance of the work for the field. As a whole, the series will provide a valuable resource both for newcomers to the research fields described, and for other scientists seeking detailed background information on special questions. Finally, it provides an accredited documentation of the valuable contributions made by today's younger generation of scientists.

Theses are accepted into the series by invited nomination only and must fulfill all of the following criteria

- They must be written in good English.
- The topic should fall within the confines of Chemistry, Physics, Earth Sciences, Engineering and related interdisciplinary fields such as Materials, Nanoscience, Chemical Engineering, Complex Systems and Biophysics.
- The work reported in the thesis must represent a significant scientific advance.
- If the thesis includes previously published material, permission to reproduce this must be gained from the respective copyright holder.
- They must have been examined and passed during the 12 months prior to nomination.
- Each thesis should include a foreword by the supervisor outlining the significance of its content.
- The theses should have a clearly defined structure including an introduction accessible to scientists not expert in that particular field.

More information about this series at http://www.springer.com/series/8790

Kadir Utku Can

Electromagnetic Form Factors of Charmed Baryons in Lattice QCD

Doctoral Thesis accepted by
the Tokyo Institute of Technology, Tokyo, Japan

Author
Dr. Kadir Utku Can
Strangeness Nuclear Physics Laboratory,
Nishina Center
RIKEN
Wako, Saitama
Japan

Supervisor
Prof. Makoto Oka
Advanced Science Research Center
Japan Atomic Energy Agency
Tokai, Ibaraki
Japan

ISSN 2190-5053 ISSN 2190-5061 (electronic)
Springer Theses
ISBN 978-981-10-8994-7 ISBN 978-981-10-8995-4 (eBook)
https://doi.org/10.1007/978-981-10-8995-4

Library of Congress Control Number: 2018936651

Printed on acid-free paper

This Springer imprint is published by the registered company Springer Nature Singapore Pte Ltd.
part of Springer Nature
The registered company address is: 152 Beach Road, #21-01/04 Gateway East, Singapore 189721, Singapore

Supervisor's Foreword

Our universe is composed of atoms, which are made of a nucleus and electrons circulating around the nucleus. The nucleus is further a composite object of elementary particles called hadrons. The most popular hadron is the proton, the nucleus of the hydrogen atom. Heavier nuclei contain multiple protons and also some neutrons, which are the neutral sibling of the proton. The last half of the twentieth century was an era for discoveries of new particles beyond the electron, proton, and neutron, which lead to the standard model of elementary particles. There the quark and gluon are introduced to describe the structure and dynamics of the hadrons (and thus the nucleus).

Analyzing the hadron structure in terms of quarks and gluons is one of the main subjects in the hadron physics. The fundamental theory of their dynamics is known as the quantum chromodynamics (QCD), which is a part of the standard model. According to QCD, quarks and gluons interact with each other through color-dependent force or the color–gauge interaction. Although QCD is as simple as expressed by a single-line Lagrangian, its outcome has full of surprises. At low energy, the color-coupling strength is so strong that the quark and the gluon cannot be isolated, but instead are completely confined in color neutral (singlet) hadrons. Furthermore, light quarks get dynamical masses due to the (chiral symmetry breaking) quark condensate in the strongly correlated vacuum. These features of the strong interaction make the analysis of QCD and comparison with experiment quite complicated.

In this thesis work, the author, Kadir Utku Can, studies the hadron structure from QCD, focusing on the baryons that contain at least one charm quark (charmed baryons). Utku's work was motivated by recent rapid developments of experimental studies in the spectroscopy of heavy-quark (charm and bottom) hadrons at various high-energy particle accelerators. In 2003, the Belle detector experimental group at KEK, Japan, found a new resonance state in the region of the charmonium spectrum. This state happens to be quite exotic; *i.e.*, it cannot simply be a charm–anti-charm bound state, but it should contain more quarks. This was followed by many other candidates of "exotic" hidden-charm multi-quark states. The theorists started to reconsider a simple picture of charmonium states and have developed new ideas, such as di-quark correlation, hadron molecule, strong channel coupling.

What is special in heavy quarks? The intriguing aspect is that the inner dynamics of charmed baryons is supposed to be distinguished from their light counterparts, *i.e.*, that of the nucleon. This is caused by the large mass of the charm quark and the asymptotic freedom, the key property of QCD. In order to investigate such properties, Utku calculated the electromagnetic form factors of the lowest lying charmed baryons. The form factors can represent, after the Fourier transform, the charge densities coming from the quarks of each flavor.

As the hadrons are formed at the low-energy regime of QCD, the effective coupling constant is significantly large and perturbative approaches are inapplicable. Then, a state-of-the-art numerical method, called lattice QCD, was applied to this study, which is the most powerful, ab initio, and non-perturbative method of QCD. The lattice QCD is a method in which quarks and gluons are regarded as dynamical variables defined on a discretized space-time lattice. They move around on the lattice according to the dynamics of QCD and form bound states, *i.e.*, hadrons. Their motion is described as a multi-dimensional path integral, which is evaluated numerically by an importance sampling method. After 30 years of its proposal, the technique has been advanced with the help of development of fast and large supercomputers. Recent calculations have achieved good accuracy that can be compared with available experimental data.

Utku has applied the lattice QCD technique to calculate the mass and form factors of heavy baryons. The thesis is based on series of papers which Utku and his collaborators published in these several years. The work required two important developments, (1) to properly implement the charm quark on the lattice and (2) to calculate three-point (vertex) functions with finite momenta. The combination of these two components has been achieved for the first time in this thesis work. He successfully obtained the form factors of the charmed baryons and estimated the distribution and the radius (extension size) of each flavor of quarks. His analysis shows a systematic behavior of the quark distributions for the charmed baryons. Most notably, he shows that the expectations of the (naive) quark model and effective field theory calculations of heavy systems are qualitatively consistent with the first principle calculations. For instance, the spin-flavor correlation of quarks in the baryon agrees qualitatively to simple (nonrelativistic) quark models. On the other hand, the quantitative results often deviate from the model calculations. This thesis contains the details of his analyses of the lattice QCD results. His results also provide insights into the dynamics of yet unobserved charmed baryon systems.

This is the first step to investigate the charmed (and in future also bottomed) hadron structure from the first principle. Utku plans to extend his calculation of the form factors further for the so-called heavy exotic hadrons, which are regarded as multi-quark resonance states.

Kadir Utku Can came from Turkey and became a graduate student at Tokyo Institute of Technology (Tokyo Tech.) under my supervision. His former supervisor, Güray Erkol, Professor at Ozyegin Univ., Istanbul, is also a long-term collaborator of mine, since he was a JSPS Postdoctoral Fellow at Tokyo Tech.

Toru Takahashi, Professor at the National Institute of Technology, Gunma College, is another key collaborator of this work. Utku's study at Tokyo Tech. was based on the collaboration between Turkey and Japan and has been very fruitful. We thank all the collaborators of our work and hope to continue the fruitful collaboration in future.

Tokai, Japan
November 2017

Prof. Makoto Oka

Acknowledgements

First and foremost, I would like to thank Prof. Makoto Oka for providing a wonderful and nurturing research environment. Without the freedom and support he had given, this work would not have been possible. I count myself privileged for being a student of his and sharing a glimpse of his knowledge. I hope one day I can step into his shoes to carry the flame further.

An exceptional person for whom I have the utmost appreciation and gratitude is Prof. Güray Erkol. Without his support, mentorship, and friendship, I would not have made it this far but be lost. I am eternally grateful to him and am proud to become a colleague of him. I cannot thank you enough, but again: *Sağolun Hocam!*.

My special thanks to Prof. Toru T. Takahashi for his support and all the discussions we had. His insight has always been a vital ingredient. It has been a lively and joyful experience in this research group thanks to two of my roommates Yokota and Iwasaki. A second thanks to Yokota-sama for being there whenever I failed to communicate in Japanese and needed help. I appreciate the friendships of Sergio, Yoshida, and Araki very much and thank the rest of the group members for the discussions and the help they have provided. Heartfelt thanks to all my friends from Turkey and Japan for creating an active social environment. They are unseen but unforgotten.

My final regards and appreciation go to my beloved family. Thank you for your endless support and encouragement. Although light-years away, our bonds kept me going with you always by my side. This whole work is dedicated to you, the greatest of all grandparents and mothers. *İ yi ki varsınız, hepinizi çok seviyorum.*

Contents

Part I
Introductory Review

Chapter 1
Introduction

Abstract In this chapter, we outline the basics of modern particle physics. We begin with briefly discussing the elementary particles and their interactions in the context of Standard model. A special attention is given to the strong force and its relation to the hadron formation along with the intriguing questions it raises. We, then, shift our attention to introducing some differing aspects of heavy-flavored hadrons with respect to lighter hadron and review the theoretical tools that are mainly used to study hadron phenomenology. At the close of the chapter, an outline of the thesis is provided.

Keywords Standard model · Strong force · Hadrons and heavy-flavored hadrons Theoretical methods

1.1 Mini Review of the Standard Model

Nature, according to humanity's accumulated knowledge so far, is governed by four forces: electromagnetic, weak, strong and gravitational force. First three of these forces have well established quantum theories while the gravitation still remains as a classical force in our toolbox of formulations. Modern understanding of the quantum theories is that, forces are mediated by the particles associated with that particular theory. For instance, photons transmit the electromagnetic force while gluons are the transmitters of the strong force. The Standard Model (SM) of particle physics combines the quantum theories for these three forces under a framework, using which we are able to calculate the interactions amongst the particles.

Elementary particle zoo of the standard model is categorised into fermions and bosons with respect to the spins of the particles. Fermions have half-integer spins (e.g. 1/2, 3/2 ...) and the bosons have integer spins (e.g. 0, 1, ...). Fermions are further classified into leptons and quarks. Leptons consist of electrically charged electron (e^-), its two heavier cousins, muon (μ^-) and tau (τ^-) and their antiparticles. In addition to these charged ones, leptons also consist of chargeless neutrinos (and anti neutrinos) of electronic (ν_e), muonic (ν_μ) and tauic (ν_τ) type. Quark sector is composed of six quarks with flavors [up (u), down (d), strange (s), charm

K. U. Can, *Electromagnetic Form Factors of Charmed Baryons in Lattice QCD*,
Springer Theses, https://doi.org/10.1007/978-981-10-8995-4_1

Table 1.1 Elementary particles of the Standard Model. PDG mass averages, electric charges in units of elementary charge *e* and spins are indicated

Fermions						Bosons			
u	$\sim$ 2.2 MeV	c	$\sim$ 1.27 GeV	t	$\sim$ 173 GeV	g	0	H	$\sim$ 125 GeV
	+2/3		+2/3		+2/3		0		0
	1/2		1/2		1/2		1		0
d	$\sim$ 4.7 MeV	s	$\sim$ 96 MeV	b	$\sim$ 4.18 GeV	γ	0		Mass
	$-1/3$		$-1/3$		$-1/3$		0		Charge
	1/2		1/2		1/2		1		Spin
e	0.511 MeV	μ	105.66 MeV	τ	1.776 GeV	Z	91.2 GeV		
	-1		-1		-1		0		
	1/2		1/2		1/2		1		
ν_e	< 2.2 eV	ν_μ	< 0.19 MeV	ν_τ	< 18.2 MeV	W	80.4 GeV		
	0		0		0		± 1		
	1/2		1/2		1/2		1		

(c), bottom (b), top (t)] and fractional charges of (2/3, -1/3, 2/3, -1/3, 2,3, -1/3)e, respectively, all having three different colors. Quarks (and gluons) carry one more quantum number, the color quantum number or, in analogy to the electric charge, color charge, which is a distinctive feature of the strong interactions. Boson sector consists of *force carrier* particles. Photon (γ) is the mediator of electromagnetic interactions with a long interaction range. Any particle that has an electric charge interacts with photon and vice versa. The $W^{\pm}$ and Z bosons carry the weak force, which is involved in the decays of radioactive particles with a short-range around the size of a hadron. Gluon (g) is associated with the strong force, the force that binds the (anti-)quarks into protons and neutrons, which in turn, are again held together by the strong force to build up the nuclei of the atoms that populate the periodic table. It has two color components. Finally, the Higgs (H) boson is an integral part of the mass generating mechanism of the SM via spontaneous symmetry breaking, in which every massive elementary particle (fermion or boson and including itself) gains its mass by interacting with it. Elementary particles are tabulated in Table 1.1.

Each interaction is described by a gauge theory with its corresponding gauge group. The total symmetry gauge group of the SM is $SU(3)_c \times SU(2) \times U(1)$ associated with the strong, weak and electromagnetic interactions respectively. The latter two interactions are unified into the electroweak theory [1–3], with its symmetry group spontaneously broken to $SU(2)_L \times U(1)_Y$ by the Higgs mechanism [4, 5] where the $W^{\pm}$ and Z bosons acquire mass and the photon remains massless. W and Z bosons have been discovered in the early eighties [6] and a long anticipated discovery of a Higgs boson, which is highly associated with the SM Higgs [7], came in 2012 by the ATLAS [8] and CMS Collaborations [9] at LHC. Electroweak physics is a vast field of subject with current studies focusing on the precision tests of theoretical predictions in the search of new physics (see Sect. 10 of Ref. [10] and references therein).

1.2 Strong Force and Hadrons

Strong sector of SM, associated with the $SU(3)_{color}$ gauge group and described by Quantum Chromodynamics (QCD), has different characteristics such that the nature of the interactions changes with the changing energy scale in a way unlike the electromagnetic or weak interactions. On one hand—at high-energy end—elementary particles interact weakly (in strength), similar to what we are familiar with from other interactions, while on the other—at low-energy end—interaction strength grows so strong that they are bound into composite particles. While the high-energy interactions can be calculated by perturbative methods, low-energy regime dynamics turns out to be challenging, rendering a perturbative approach (in powers of the strong coupling) ineffective.

Low-energy dynamics of QCD leads to the formation of hadrons—color neutral combinations of quarks and gluons. In the simplest picture, depending on the quark composition, they are divided into two categories: Mesons, composed of a quark-anti quark pair and baryons, composed of three quarks. In addition to these simple formations, hadrons might be composed of multi quarks and gluons, or form molecular states. Such states are called exotic states and after decades of elusive searches, recent experimental evidence on exotic mesons (see mini review on pentaquarks in Ref. [10]) and pentaquark candidates [11] looks promising to improve our understanding of hadron formation.

Early era of hadron physics is involved in the classification of the emerging hadrons, shaping our knowledge on the strong force to its current form. Concepts of quarks, color quantum number and its association with the $SU(3)$ group, flavor symmetries for hadron categorisation and all emerged before a field theoretic description of QCD paving the way for an enhanced understanding. A milestone is the Eightfold Way classification. It is a classification scheme devised independently by Gell-Mann [12] and Ne'emann [13] in 1961 based on representations of the $SU(3)$ group with an underlying flavor symmetry—note that the three stands for the three flavors of quarks, namely u, d and s, not the color and assumes all have the same mass—which classified the eight known baryons at that time. It has also provided a mass formula between the baryons, Gell-Mann-Okubo relation [14], which was satisfied by the experimental masses of the observed baryons but also has been used to predict the mass of the, yet unobserved, Ω^- particle. Observation of the Ω^- particle close to its predicted mass [15] is a notable achievement of the classification scheme. Classification based on flavor symmetries is a crucial part of hadron physics in which the quantum numbers of unobserved hadrons or the undetermined ones of known hadrons are usually assigned according to group theoretical foundations. Flavor symmetries are not exact however. We know that the flavor $SU(3)$ symmetry is broken mildly and the flavor $SU(4)$ symmetry, incorporating the charm quark, is broken badly due to the heavy mass of the charm. Yet, there is satisfactory evidence that expectations hold and they provide us with invaluable guidance in calculations.

Naturally, with so many around with rich spectra, we are interested in the masses of hadrons. One intriguing aspect, for example, is how to calculate the mass of a hadron from QCD. Take proton for instance. It is situated at the core of the most abundant element in the visible Universe and consists of two up and one down valence quarks which have masses of only a few MeV compared to the almost 1 GeV mass of the proton. Remaining part is the binding energy due to the complex quark-gluon dynamics. Higgs mechanism provides the masses of the quarks for sure, but, strictly speaking, the mass of the visible Universe is dynamically generated via strong interactions. Yet, even though we have a description, we are unable to calculate the mass of the proton—by a perturbative expansion in powers of the strong coupling at least.

It is another challenge to understand the structure of the hadrons. Being composite objects, we are interested in how the components are distributed inside or how they combine to give the hadron properties. Following on, we can immediately come up with questions like: how is the electric charge distributed? Does any charge-neutral hadron have a charged core in analogy to an atom or is the charge distributed homogeneously? What about the magnetic structure? Does the magnetisation behave similar to the electric charge density? Are there visible effects of the magnetic moments? Can we resolve the spin structure of a hadron? and so on... For instance, consider the electric charge radius of the proton: There is a $\sim 7\sigma$ discrepancy between its electron based (e-p scattering and hydrogen spectroscopy) measured value ($R_E^e = 0.8775(71)$ fm) [16] and the one extracted via muonic hydrogen experiments ($R_E^\mu = 0.84169(66)$ fm) [17]. We know that the baryons are composed of three quarks and a sea of quarks-anti quarks and gluons. But consider for instance, how much of the spin do these constituents share? Deep-Inelastic scattering experiments indicate that only 1/3 of the proton's spin is carried by quarks and anti-quarks [18, 19]. These two simple examples show that there are still unresolved issues in our understanding of the inner dynamics of the hadrons, which, in principle, can be addressed by QCD calculations. Usually the examples and motivations are related to the nucleon due to the historical development and bulk of experimental results, however, the same questions and motivations apply to all hadrons and heavier sectors come with their own questions.

1.3 Heavy-Flavored Hadrons

Heavy sector, in the general sense, corresponds to the hadrons that have at least one charm or bottom quark as a valence quark but in the context of this thesis we will restrict our attention to charmed hadrons only. Discovery of the charm quark dates back to the observation of heavy meson resonances in e^+e^- annihilation experiments conducted at the BNL [20] and SLAC [21] in the seventies. These resonances are interpreted as bound states of a charm-anti charm pair, or charmonium in general, and the ground state is now known as the J/ψ meson while the excited states are named only as the ψ. There are now many identified charmed mesons, either charmonium ($c\bar{c}$) or open-charmed with one charm quark such as $D^{(*)+,0}$ ($c\bar{d}$, $c\bar{u}$) or $D_s^{(*)\pm}$ ($c\bar{s}$, $\bar{c}s$) mesons and their excited states.

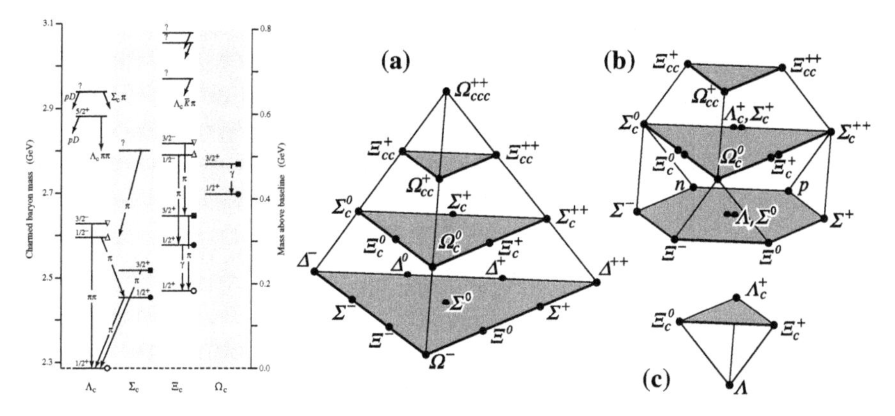

Fig. 1.1 Current experimental situation of the charmed baryons (left). $SU(4)$ classification of spin-1/2 (**b**, **c**) and spin-3/2 (**a**) baryons (right). Base levels contain the $SU(3)$ multiplets. Both figures are taken from Ref. [10]

Experimental information on the baryon sector is somewhat limited however. As of today, there are only 18 observed and established charmed-baryon states all of which contain only one charm quark and the spin-parity values of only a few has been experimentally measured. Left panel of Fig. 1.1 depicts the current experimental situation with four species of singly-charmed baryons and their excited states. Existence of the rest of the charmed baryons are inferred from the $SU(4)$ flavor symmetry arguments with assigned spin-parity values of some with respect to theory calculations. $SU(4)$ multiplets containing positive parity spin-1/2 and spin-3/2 baryons consisting of u, d, s and c quarks are shown in the right panel of Fig. 1.1 for reference. SELEX Collaboration claimed to have observed the doubly-charmed Ξ_{cc} baryon [22, 23], however that has been the only instance so far and the searches in the BaBar [24] and Belle [25] experiments did not confirm the claims of SELEX although they had better statistics.

An interesting aspect of heavy quarks is the change of dynamics as $m_Q \to \infty$, known as the heavy-quark symmetry (HQS). As a consequence, spin of the heavy quark decouples from the system since it is proportional to $1/m_Q$ and the system is characterised by the light degrees of freedom. One therefore would expect the effects of hyperfine splitting to diminish as $m_Q \to \infty$ such that the masses of the particles with the same quark content but different total spins would get closer. Such a trend is evident in the meson, $m_{J=1} - m_{J=0}$, and baryon, $m_{J=3/2} - m_{J=1/2}$, mass splittings given below (evaluated using the PDG averages) for some illustrative hadrons,

$$\begin{aligned}
m_\rho - m_\pi &= 630\,\text{MeV}, & m_\Delta - m_N &= 290\,\text{MeV},\\
m_{K^*} - m_K &= 390\,\text{MeV}, & m_{\Sigma^*} - m_\Sigma &= 195\,\text{MeV},\\
m_{D^*} - m_D &= 140\,\text{MeV}, & m_{\Sigma_c^*} - m_{\Sigma_c} &= 65\,\text{MeV},\\
m_{B^*} - m_B &= 45\,\text{MeV}, & m_{\Sigma_b^*} - m_{\Sigma_b} &= 20\,\text{MeV}.
\end{aligned}$$

Effects of the HQS can be identified in structural observables as well. For instance, in Ref. [26], we investigate the electromagnetic structure of D-mesons with varying light-quark masses. Electromagnetic observables tend to coincide as the light quark gets heavier, which one would expect since the effect of the spin-spin interaction component of the form $(\vec{\sigma}_Q \cdot \vec{\sigma}_q / m_Q m_q)$ would decrease with increasing m_q.

Forthcoming experiments with a heavy-hadron physics programme at major experimental facilities, e.g. J-PARC, SuperKEKB, BES-III etc., are expected to provide a wealth of information which calls for a better understanding of the heavy-sector dynamics from theoretical grounds. Exotic state candidates mainly observed in the heavy sector also urges us to improve our understanding of the heavy quark dynamics. However, since perturbative QCD breaks down at the typical hadronic energy scale, we have to resort to methods that can incorporate the non-perturbative effects—be it rather naive model-dependent approaches or state-of-the-art ab initio calculations.

1.4 Contents of the Toolbox

Preceding and following the formulation of QCD there have been and are several methods to study its low-energy phenomenology such as quark models, where one considers constituent quarks—an object that implicitly include the dynamical effects of valence quarks, sea quarks and gluons—as interacting effective degrees of freedom of the hadrons to describe the interactions under a confining potential. There are many quark models falling into different categories starting from simple, non-relativistic formulations to more involved ones taking into consideration dynamics like relativistic corrections, spin-spin or spin-orbit interactions etc. One then proceeds by solving the Schrödinger equation with a choice of a proper confining potential and extracts the hadron properties. Spectroscopy and electromagnetic properties of charmed baryons, in particular, are studied in many variants of quark models [27–36], which we compare our relevant results in the appropriate sections.

Effective field theory approach, which respects the underlying symmetries of QCD but replaces the elementary ((anti-)quark, gluon) degrees of freedom by effective ones (mesons and baryons) and reformulates QCD to account for the non-perturbative effects, provides a systematic framework to study low-energy QCD phenomena. Chiral perturbation theory [37, 38], for instance, is one successful example where one writes down the most general Lagrangian containing all the effective interactions consistent with the underlying symmetry principles to study low-energy dynamics by a perturbative expansion in the order parameter of the theory. Chiral symmetry, which arises in the chiral, $m_q \to 0$, limit, and its spontaneous breaking, leading to the emergence of Nambu-Goldstone bosons—pion being the lightest of them—underlies the chiral perturbation theory. Being an effective theory, however, it has an applicable energy range, $m_q \ll \Lambda_{QCD}$, and naturally breaks down when one goes out of bounds. It has been successfully applied in the vicinity of light quark energies [39] although their small mass breaks the chiral symmetry explicitly they are still small enough to be neglected. Expansion already exhibits a poor convergence when one adds the strange quark the mass of which is close to Λ_{QCD}, nevertheless there are remedies to this situation (see Ref. [40]). Another limitation is related to the so-called low-energy constants associated with each term in the expansion (interaction vertices), which are input parameters in need of a determination by other methods, *i.e.* fitting to experimental data or calculating via different methods. Such shortcomings, however, do not diminish its importance as a useful tool where there is a close cooperation between the chiral perturbation theory and lattice QCD since former provides a much insightful description for the quark-mass dependence of observables and systematic effects such as the finite volume and discretisation [41–43] while the latter helps with determining the low-energy constants needed for the chiral expansion.

On the other edge of the scale, namely in the $m_q \to \infty$ limit, one can exploit the HQS and devise a scheme based on a combination of Heavy Quark Expansion (HQE) and Operator Product Expansion (OPE) [44] to account for the non-perturbative effects. HQE provides an order parameter of the form $1/m_Q$ for the expansion and OPE incorporates the short- and long-distance behaviour in terms of the so-called Wilson coefficients (calculable perturbatively) and the expectation values of local operators of dimension $d \geq 4$ constructed from quark bilinears, respectively. Basically, the QCD Lagrangian is decomposed into two parts, $\mathcal{L}_{QCD}(\mu) = \mathcal{L}_{\text{light}}(\mu) + \mathcal{L}_{\text{Heavy}}(\mu)$, where the light part is the usual QCD Lagrangian but contains the dynamics of the light quarks, $m_q \ll m_c$, only and treated perturbatively while the heavy part is expressed in terms of non-relativistic quark fields and takes a general form as $\mathcal{L}_{\text{Heavy}}(\mu) = \mathcal{O}^{(4)} + c^{(5)}(\mu)\mathcal{O}^{(5)}/m_Q + c^{(6)}(\mu)\mathcal{O}^{(6)}/m_Q^2 + \mathcal{O}(1/m_Q^3)$, where $c^{(d)}$ and $\mathcal{O}^{(d)}$ are Wilson coefficients and local operators of dimension d. Decomposing essentially introduces an additional energy scale, μ, which defines an applicability range for the prescription, $\Lambda_{QCD} \ll \mu \ll m_Q$, where the short-distance effects can be described perturbatively in powers of strong coupling constant, $\alpha_s(\mu)$, and the long-distance dynamics exhibit non-perturbative effects. A typical choice is $\mu \sim 1$ GeV which is plausible for investigating b-hadron properties since $\mu \sim 1\text{GeV} \ll m_b$ but the proximity of the $m_c = 1.2$ GeV to the choice of $\mu \sim 1$ GeV might make such an approach tricky.

QCD Sum Rules [45, 46] is another example of a non-perturbative method but differs from the previous ones in the sense that it is based on the elementary degrees of freedom of QCD. It is a widely used, powerful method providing valuable insight into hadron properties [47, 48]. In this approach one calculates the vacuum expectation values, correlation functions, of observables in question from two different scales of the theory where an OPE in terms of quark and gluon condensates relates the elementary d.o.f. to the hadronic observables defined in the correlation function. There are, however, certain limitations stemming from approximations such as the truncation of the OPE at some order or introduction of artificial quantities, which leads to an inherit uncertainty of $\sim$10–20%. QCDSR literature is very rich in content with applications including spectroscopy of light, heavy and possible exotic hadrons, form factor calculations, finite temperature and density investigations and so on. Previous studies on charmed baryons had a focus on their spectrum [49, 50] and magnetic moments [51, 52].

Lattice approach has its roots in the seventies [53] and one might say it has surpassed its predecessors in its current state-of-the-art form in some aspects thanks to the evolution of the computer technology and significant improvement efforts on numerical algorithms. It, of course, has its own caveats. There are fields where a lattice application suffers from technicalities, e.g. the sign problem in finite density calculations, or the need to control the systematic uncertainties arising due to the formulation itself. In principle, however, lattice approach is ab initio, free from any assumptions or model-dependencies in the QCD Lagrangian level and a systematic study is possible to control the errors with current calculations reaching percent, even sub-percent, level accuracy challenging the predictions of the Standard Model [54–56]. In a nutshell, one discretises the space-time continuum and reformulates the QCD Lagrangian in a Euclidean setup to compute the correlation functions defined in a Feynman path integral formalism via numerical Monte Carlo methods. There is a close connection between the lattice approach and the ideas of statistical mechanics where one obtains the observables by basically averaging over QCD vacuum configurations. Since its advent, lattice methods have been applied to many sub-fields of QCD. Currently, it is possible to make fully dynamical simulations with physical or almost-physical quark masses and controlled extrapolations to infinite-volume and continuum limit. From a hadron physics point of view, most of the modern efforts converge on the spectroscopy of light and heavy hadrons and hadron-hadron interactions [57] and structure calculations of light pseudoscalar and vector mesons and nucleon [58]. Form factors of low-lying octet [59–61] and decuplet [62–64] baryons are also studied. Lattice QCD calculations in the charmed baryon sector however is limited to spectroscopy calculations [42, 65–67] only. Results on charmed baryon electromagnetic form factors and observables that we present here have the characteristic of being the first calculations in a lattice QCD framework. It is worth mentioning that, parallel to the calculations presented here, we have a continuing program on electromagnetic and axial transitions of charmed baryons—results of which are out of the scope of this work but already available in Refs. [68, 69].

1.5 Outline of the Thesis

We organise this thesis as follows: In Chap. 2, we give the QCD formalism and discuss the hadron structure by a brief historical account. Chapter 3 focuses on the lattice approach, where we investigate its implementation in detail. We discuss the discretisation of the space-time and the QCD action and sketch its application. Chapter 4 consists of the technical details of our calculations. We show how to extract the masses and calculate the electromagnetic form factors of spin-1/2 and spin-3/2 baryons in a general formalism followed by the details of our simulation setup. Our results on the electromagnetic observables of charmed baryons are presented and discussed in detail along with comparisons to the light sector and results of the other methods in Chap. 5. Conclusions and a summary of our work is given in Chap. 6.

References

1. S.L. Glashow, Partial-symmetries of weak interactions. Nucl. Phys. 22(4), 579–588 (1961). ISSN 0029-5582. https://doi.org/10.1016/0029-5582(61)90469-2, http://www.sciencedirect.com/science/article/pii/0029558261904692
2. S. Weinberg, A model of leptons. Phys. Rev. Lett. **19**, 1264–1266 (1967). https://doi.org/10.1103/PhysRevLett.19.1264
3. A. Salam, Weak and electromagnetic interactions. Conf. Proc. **C680519**, 367–377 (1968)
4. P.W. Higgs, Broken symmetries, massless particles and gauge fields. Phys. Lett. 12(2), 132–133 (1964). ISSN 0031-9163. https://doi.org/10.1016/0031-9163(64)91136-9, http://www.sciencedirect.com/science/article/pii/0031916364911369
5. F. Englert, R. Brout, Broken symmetry and the mass of gauge vector mesons. Phys. Rev. Lett. **13**, 321–323 (1964). https://doi.org/10.1103/PhysRevLett.13.321
6. G. Arnison et al., Experimental observation of events with large missing transverse energy accompanied by a jet or a photon(s) in p anti-p collisions at s**(1/2)=540-GeV. Phys. Lett. B **139**, 115 (1984). https://doi.org/10.1016/0370-2693(84)90046-7
7. J. Ellis, T. You. Updated global analysis of Higgs couplings. J. High Energy Phys. 2013(6), 103 (2013). ISSN 1029-8479. https://doi.org/10.1007/JHEP06(2013)103
8. G. Aad et al., Observation of a new particle in the search for the standard model Higgs boson with the ATLAS detector at the LHC. Phys. Lett. B **716**, 1–29 (2012). https://doi.org/10.1016/j.physletb.2012.08.020
9. S. Chatrchyan et al., Observation of a new boson at a mass of 125 GeV with the CMS experiment at the LHC. Phys. Lett. B **716**, 30–61 (2012). https://doi.org/10.1016/j.physletb.2012.08.021
10. C. Patrignani et al., Review of Particle Physics. Chin. Phys. **C40**(10), 100001 (2016). https://doi.org/10.1088/1674-1137/40/10/100001
11. R. Aaij et al., Observation of $J/\psi p$ resonances consistent with Pentaquark States in $\Lambda_b^0 \to J/\psi K^- p$ Decays. Phys. Rev. Lett. **115**, 072001 (2015). https://doi.org/10.1103/PhysRevLett.115.072001
12. M. Gell-Mann, The Eightfold Way: A Theory of Strong Interaction Symmetry, (1961)
13. Y. Ne'eman. Derivation of strong interactions from a gauge invariance. Nucl. Phys. 26(2), 222–229 (1961). ISSN 0029-5582. https://doi.org/10.1016/0029-5582(61)90134-1, http://www.sciencedirect.com/science/article/pii/0029558261901341
14. S. Okubo, Note on unitary symmetry in strong interactions. Prog. Theor. Phys. **27**(5), 949–966 (1962). https://doi.org/10.1143/PTP.27.949., http://ptp.oxfordjournals.org/content/27/5/949.abstract
15. V.E. Barnes et al., Observation of a hyperon with strangeness minus three. Phys. Rev. Lett. **12**, 204–206 (1964). https://doi.org/10.1103/PhysRevLett.12.204.
16. P.J. Mohr, B.N. Taylor, D.B. Newell, Codata recommended values of the fundamental physical constants: 2010*. Rev. Mod. Phys. **84**, 1527–1605 (2012). https://doi.org/10.1103/RevModPhys.84.1527
17. A. Antognini et al., Proton structure from the measurement of 2s-2p transition frequencies of muonic hydrogen. Science, **339**(6118), 417–420 (2013). ISSN 0036-8075. https://doi.org/10.1126/science.1230016, http://science.sciencemag.org/content/339/6118/417
18. A. Airapetian et al., Precise determination of the spin structure function g(1) of the proton, deuteron and neutron. Phys. Rev. D **75**, 012007 (2007). https://doi.org/10.1103/PhysRevD.75.012007
19. VYu. Alexakhin et al., The Deuteron Spin-dependent Structure Function g1(d) and its First Moment. Phys. Lett. B **647**, 8–17 (2007). https://doi.org/10.1016/j.physletb.2006.12.076
20. J.J. Aubert et al., Experimental observation of a heavy particle j. Phys. Rev. Lett. **33**, 1404–1406 (1974). https://doi.org/10.1103/PhysRevLett.33.1404
21. J.E. Augustin et al., Discovery of a narrow resonance in e^+e^- annihilation. Phys. Rev. Lett. **33**, 1406–1408 (1974). https://doi.org/10.1103/PhysRevLett.33.1406

22. M. Mattson et al., First observation of the doubly charmed baryon Xi+(cc). Phys. Rev. Lett. **89**, 112001 (2002). https://doi.org/10.1103/PhysRevLett.89.112001
23. A. Ocherashvili et al., Confirmation of the double charm baryon Xi+(cc)(3520) via its decay to p D+ K-. Phys. Lett. B **628**, 18–24 (2005). https://doi.org/10.1016/j.physletb.2005.09.043
24. B. Aubert et al., Search for doubly charmed baryons Xi(cc)+ and Xi(cc)++ in BABAR. Phys. Rev. D **74**, 011103 (2006). https://doi.org/10.1103/PhysRevD.74.011103
25. R. Chistov et al., Observation of new states decaying into Lambda(c)+ K- pi+ and Lambda(c)+ K0(S) pi-. Phys. Rev. Lett. **97**, 162001 (2006). https://doi.org/10.1103/PhysRevLett.97.162001
26. K.U. Can, G. Erkol, M. Oka, A. Ozpineci, T.T. Takahashi, Vector and axial-vector couplings of D and D* mesons in $2 + 1$ flavor lattice QCD. Phys. Lett. B **719**(1–3), 103–109 (2013). ISSN 0370-2693. https://doi.org/10.1016/j.physletb.2012.12.050, http://www.sciencedirect.com/science/article/pii/S0370269312013032
27. A.P. Martynenko, Ground-state triply and doubly heavy baryons in a relativistic three-quark model. Phys. Lett. B **663**, 317–321 (2008). https://doi.org/10.1016/j.physletb.2008.04.030
28. W. Roberts, Muslema Pervin, Heavy baryons in a quark model. Int. J. Mod. Phys. A **23**, 2817–2860 (2008). https://doi.org/10.1142/S0217751X08041219
29. B. Julia-Diaz, D.O. Riska, Baryon magnetic moments in relativistic quark models. Nucl. Phys. A **739**, 69–88 (2004). https://doi.org/10.1016/j.nuclphysa.2004.03.078
30. A. Faessler, T. Gutsche, M.A. Ivanov, V.E. Lyubovitskij, J.G. Korner et al., Magnetic moments of heavy baryons in the relativistic three-quark model. Phys. Rev. D **73**, 094013 (2006). https://doi.org/10.1103/PhysRevD.73.094013
31. C. Albertus, E. Hernandez, J. Nieves, J.M. Verde-Velasco, Static properties and semileptonic decays of doubly heavy baryons in a nonrelativistic quark model. Eur. Phys. J. A **32**, 183–199 (2007). https://doi.org/10.1140/epja/i2007-10364-y, 10.1140/epja/i2008-10547-0
32. N. Sharma, H. Dahiya, P.K. Chatley, M. Gupta, Spin $1/2^+$, spin $3/2^+$ and transition magnetic moments of low lying and charmed baryons. Phys. Rev. D **81**, 073001 (2010). https://doi.org/10.1103/PhysRevD.81.073001
33. N. Barik, M. Das, Magnetic moments of confined quarks and baryons in an independent-quark model based on Dirac equation with power-law potential. Phys. Rev. D **28**, 2823–2829 (1983). https://doi.org/10.1103/PhysRevD.28.2823
34. S. Kumar, R. Dhir, R.C. Verma, Magnetic moments of charm baryons using effective mass and screened charge of quarks. J. Phys. **G31**, 141–147 (2005). https://doi.org/10.1088/0954-3899/31/2/006
35. B. Patel, A.K. Rai, P.C Vinodkumar, Masses and magnetic moments of heavy flavour baryons in hyper central model. J. Phys. **G35**, 065001 (2008). https://doi.org/10.1088/1742-6596/110/12/122010, https://doi.org/10.1088/0954-3899/35/6/065001
36. T. Yoshida, E. Hiyama, A. Hosaka, M. Oka, K. Sadato, Spectrum of heavy baryons in the quark model. Phys. Rev. D **92**(11), 114029 (2015). https://doi.org/10.1103/PhysRevD.92.114029
37. J. Gasser, H. Leutwyler, Chiral perturbation theory: expansions in the mass of the strange quark. Nucl. Phys. B **250**, 465 (1985), https://doi.org/10.1016/0550-3213(85)90492-4
38. S. Weinberg, Phenomenological Lagrangians. Physica **A96**, 327 (1979)
39. J. Gasser, H. Leutwyler, Quark masses. Phys. Rep. **87**(3), 77–169 (1982). ISSN 0370-1573. https://doi.org/10.1016/0370-1573(82)90035-7, http://www.sciencedirect.com/science/article/pii/0370157382900357
40. B.C. Tiburzi, Chiral perturbation theory, in *Lattice QCD for Nuclear Physics*, ed. by L. Huey-Wen, B. Harvey Meyer (Springer International Publishing, Cham, 2015), pp. 107–152. ISBN 978-3-319-08022-2. https://doi.org/10.1007/978-3-319-08022-2_4
41. S. Aoki, K.-I. Ishikawa, N. Ishizuka, T. Izubuchi, D. Kadoh, K. Kanaya, Y. Kuramashi, Y. Namekawa, M. Okawa, Y. Taniguchi, A. Ukawa, N. Ukita, T. Yoshie, 2+1 Flavor lattice QCD toward the physical point. Phys. Rev. D **79**, 034503 (2009). https://doi.org/10.1103/PhysRevD.79.034503

42. Z.S. Brown, W. Detmold, S. Meinel, K. Orginos, Charmed bottom baryon spectroscopy from lattice QCD. Phys. Rev. D **90**(9), 094507 (2014). https://doi.org/10.1103/PhysRevD.90.094507
43. L. Liu, H.-W. Lin, K. Orginos, A. Walker-Loud, Singly and doubly charmed J=1/2 baryon spectrum from lattice QCD. Phys. Rev. D **81**, 094505 (2010). https://doi.org/10.1103/PhysRevD.81.094505
44. K.G. Wilson, Non-lagrangian models of current algebra. Phys. Rev. **179**, 1499–1512 (1969). https://doi.org/10.1103/PhysRev.179.1499
45. M.A. Shifman, A.I. Vainshtein, V.I. Zakharov, Qcd and resonance physics. Theoretical foundations. Nucl. Phys. B **147**(5), 385–447 (1979a). ISSN 0550-3213. https://doi.org/10.1016/0550-3213(79)90022-1, http://www.sciencedirect.com/science/article/pii/0550321379900221
46. M.A. Shifman, A.I. Vainshtein, V.I. Zakharov. Qcd and resonance physics. Applications. Nucl. Phys. B **147**(5), 448–518 (1979b). ISSN 0550-3213. https://doi.org/10.1016/0550-3213(79)90023-3, http://www.sciencedirect.com/science/article/pii/0550321379900233
47. B.L. Ioffe, QCD at low energies. Prog. Part. Nucl. Phys. **56**, 232–277 (2006). https://doi.org/10.1016/j.ppnp.2005.05.001
48. L.J. Reinders, H. Rubinstein, S. Yazaki, Hadron properties from QCD sum rules. Phys. Rep. **127**(1), 1–97 (1985). ISSN 0370-1573. https://doi.org/10.1016/0370-1573(85)90065-1
49. S. Groote, J.G. Korner, O.I. Yakovlev, QCD sum rules for heavy baryons at next-to-leading order in alpha(s). Phys. Rev. D **55**, 3016–3026 (1997)
50. J.-R. Zhang, M.-Q. Huang, Heavy baryon spectroscopy in QCD. Phys. Rev. D **78**, 094015 (2008). https://doi.org/10.1103/PhysRevD.78.094015
51. T.M. Aliev, A. Ozpineci, M. Savci, The magnetic moments of lambda(b) and lambda(c) baryons in light cone QCD sum rules. Phys. Rev. D **65**, 056008 (2002). https://doi.org/10.1103/PhysRevD.65.056008
52. S.-L. Zhu, P., W.P. Hwang, Z.-S. Yang, Σ_c and Λ_c magnetic moments from QCD spectral sum rules. Phys. Rev. D **56**, 7273–7275 (1997). https://doi.org/10.1103/PhysRevD.56.7273
53. K.G. Wilson, Confinement of quarks. Phys. Rev. D **10**, 2445–2459 (1974). https://doi.org/10.1103/PhysRevD.10.2445
54. C.M. Bouchard, Testing the standard model under the weight of heavy flavors. PoS LATTICE **2014**(002) (2015)
55. A.X. El-Khadra, Quark flavor physics review. PoS LATTICE **2013**(001) (2014)
56. S. Aoki et al., Review of lattice results concerning low-energy particle physics. Eur. Phys. J. C **74**, 2890 (2014). https://doi.org/10.1140/epjc/s10052-014-2890-7
57. S. Prelovsek, Hadron Spectroscopy. PoS LATTICE **2014**, 015 (2014)
58. M. Constantinou, Hadron structure. PoS LATTICE **2014** 001 (2015)
59. S. Boinepalli, D.B. Leinweber, A.G. Williams, J.M. Zanotti, J.B. Zhang, Precision electromagnetic structure of octet baryons in the chiral regime. Phys. Rev. D **74**, 093005 (2006). https://doi.org/10.1103/PhysRevD.74.093005
60. P.E. Shanahan, R. Horsley, Y. Nakamura, D. Pleiter, P.E.L. Rakow, G. Schierholz, H. Stuben, A.W. Thomas, R.D. Young, J.M. Zanotti, Magnetic form factors of the octet baryons from lattice QCD and chiral extrapolation. Phys. Rev. D **89**, 074511 (2014a). https://doi.org/10.1103/PhysRevD.89.074511
61. P.E. Shanahan, A.W. Thomas, R.D. Young, J.M. Zanotti, R. Horsley et al., Electric form factors of the octet baryons from lattice QCD and chiral extrapolation. Phys. Rev. D **90**, 034502 (2014b). https://doi.org/10.1103/PhysRevD.90.034502
62. S. Boinepalli, D.B. Leinweber, P.J. Moran, A.G. Williams, J.M. Zanotti, J.B. Zhang, Electromagnetic structure of decuplet baryons towards the chiral regime. Phys. Rev. D **80**, 054505 (2009). https://doi.org/10.1103/PhysRevD.80.054505.
63. C. Alexandrou, T. Korzec, G. Koutsou, Th Leontiou, C. Lorce et al., Delta-baryon electromagnetic form factors in lattice QCD. Phys. Rev. D **79**, 014507 (2009). https://doi.org/10.1103/PhysRevD.79.014507

64. C. Alexandrou, T. Korzec, G. Koutsou, J.W. Negele, Y. Proestos, The Electromagnetic form factors of the Ω^- in lattice QCD. Phys. Rev. D **82**, 034504 (2010). https://doi.org/10.1103/PhysRevD.82.034504
65. Y. Namekawa et al., Charmed baryons at the physical point in 2+1 flavor lattice QCD. Phys. Rev. D **87**(9), 094512 (2013). https://doi.org/10.1103/PhysRevD.87.094512
66. C. Alexandrou, V. Drach, K. Jansen, C. Kallidonis, G. Koutsou, Baryon spectrum with $N_f = 2+1+1$ twisted mass fermions. Phys. Rev. D **90**(7), 074501 (2014). https://doi.org/10.1103/PhysRevD.90.074501
67. R.A. Briceno, H.-W. Lin, D.R. Bolton, Charmed-baryon spectroscopy from lattice QCD with $N_f = 2+1+1$ Flavors. Phys. Rev. D **86**, 094504 (2012). https://doi.org/10.1103/PhysRevD.86.094504
68. H. Bahtiyar, K.U. Can, G. Erkol, M. Oka, $\omega_c\gamma \to \omega_c^*$ transition in lattice QCD. Phys. Lett. B **747**, 281–286 (2015). ISSN 0370-2693. https://doi.org/10.1016/j.physletb.2015.06.006, http://www.sciencedirect.com/science/article/pii/S0370269315004281
69. K.U. Can, G. Erkol, M. Oka, T.T. Takahashi, $\Lambda_c\Sigma_c\pi$ coupling and $\Sigma_c \to \Lambda_c\pi$ decay in lattice QCD (2016), arXiv:1610.09071

Chapter 2
Quantum Chromodynamics

Abstract This chapter is devoted to making the reader familiar with the ideas underlying the QCD and its formalism. We briefly discuss the advent of the color quantum number, which is a unique feature and centerpiece of strong interactions. In what follows, QCD is presented formally as a quantum field theory, where we discuss its energy scale-dependent characteristics and how it is related to the formation of hadrons. Finally, by following the early experimental developments, we give a historical account of the evidence for the hadron structure and sketch the simple formalism that is commonly used to study it in a theoretical approach.

Keywords Color charge · Quantum field theory · Strong coupling constant
Confinement · Hadron structure

2.1 Introduction

Quantum chromodynamics (QCD) is the theory governing the interactions of quarks, antiquarks and gluons. In contrast to the electromagnetic interactions, elementary particles of QCD carry one more quantum number, the color charge, which we account responsible for the strong interactions. Postulation of the existence of such an additional quantum number can be traced to the early stages of hadron spectroscopy experiments, to a pre-QCD era, accounts of which are given in textbooks to a great extent.

In Sect. 2.2, we give a brief discussion on the color quantum number for completeness. It is followed by Sect. 2.3 where we sketch the field theory formalism of QCD and discuss its two distinct phenomenon: Asymptotic freedom and confinement. An outcome of the confinement is the formation of hadrons and since they are composite objects formed by quarks, anti-quarks and gluons they have an inner-structure which can be probed by experiments or calculated by non-perturbative approaches. In Sect. 2.4 we briefly discuss the experimental evidence and introduce the form factors as well as presenting their physical interpretations.

K. U. Can, *Electromagnetic Form Factors of Charmed Baryons in Lattice QCD*,
Springer Theses, https://doi.org/10.1007/978-981-10-8995-4_2

2.2 Color Quantum Number

Necessity to introduce an additional quantum number arises due to the fact that baryons have half-integer spins and therefore they are classified as fermions being subject to the Pauli exclusion principle: Their wave functions must be *antisymmetric* under interchange of two identical quarks. Non-relativistic quark-model wave function of a baryon can be decomposed to its space, spin and flavor wave function components as,

$$\psi_{\mathcal{B}} = \psi_{space}\psi_{spin}\psi_{flavor}, \tag{2.1}$$

ignoring our knowledge of color quantum number. Baryon spectroscopy experiments in 1960s revealed the existence of Δ^{++} ($J = \frac{3}{2}$) baryon which is composed of three up quarks, uuu, leading to questioning the decomposition given above. Combination of three u-quarks is obviously flavor symmetric. Spatial wave function of a ground state Δ^{++} is symmetric since its orbital angular momentum is zero, $l = 0$, and to satisfy a spin-3/2 state spins of the quarks should be aligned with a symmetric spin wave function configuration, $|\frac{3}{2}\frac{3}{2}\rangle = (u \uparrow\ u \uparrow\ u \uparrow)$, rendering the total wave function symmetric.

Introducing the additional color degree of freedom with an antisymmetric wave function,

$$\psi_{color} = \varepsilon^{ijk}\psi^i\psi^j\psi^k, \tag{2.2}$$

where i, j and k are color indices running from 1 to 3, resolves the problem. With the addition of this extra quantum number, the wave function of a baryon in the simplest way becomes,

$$\psi_{\mathcal{B}} = \overbrace{\underbrace{\psi_{space}\psi_{spin}\psi_{flavor}}_{\text{symmetric}}\ \underbrace{\psi_{color}}_{\text{anti-symmetric}}}^{\text{antisymmetric}}, \tag{2.3}$$

satisfying the Pauli principle.

There are other experimental evidences for the existence of color as well such as an analysis of the cross-section ratios of electron-positron annihilation into hadrons and muons [1] reveals that color should play a role and there should be three different colors. Also, the consistency between the predicted and observed neutral pion decay rate is achieved only if there are three distinct colors. Historically, color quantum number is introduced by Greenberg [2] and later put into the context of a local $SU(3)_{color}$ gauge group by Nambu [3] and Han and Nambu [4].

2.3 Field Theory Formalism

Our modern understanding of the strong interactions of the elementary particles is embedded in the formulation of Quantum Chromodynamics (QCD) as a non-Abelian gauge theory with a local $SU(3)$ color symmetry. The Lagrangian density of the QCD holds the details of the interactions and is written as,

$$\mathcal{L}[\psi, \bar{\psi}, A] = \sum_q \bar{\psi}_q \left(i\mathcal{D}_\mu \gamma^\mu - m_q\right) \psi_q - \frac{1}{4} F^a_{\mu\nu} F^{\mu\nu}_a, \tag{2.4}$$

where the explicit sum over the quark flavors, $q = u, d, s, c, b, t$ in the fermionic part denotes the fact that the strong force acts similarly on the quarks independent of their flavor. Quark fields ψ_q and $\bar{\psi}_q$ carry two more indices, one denoting their spin-1/2 nature, the so-called Dirac index, running from 1 to 4 and the other, color index, running from 1 to 3 due to the fact that quarks carry color charge and come in three colors [5]. m_q is the mass of the quark q, and it is same for any of its colored versions.

$\mathcal{D}_\mu$ is the covariant derivative defined as

$$\mathcal{D}_\mu = \partial_\mu - i g_s \frac{\lambda_a}{2} A^a_\mu, \tag{2.5}$$

which follows from the invariance of the theory under local $SU(3)_c$ transformations and holds the dynamics of the quark-gluon interactions where the coupling constant g_s is related to the strong coupling constant of QCD via $g_s = \sqrt{4\pi\alpha_s}$ and A^a_μ is the gluon field. λ_a are 3×3 Gell-Mann matrices, the generators of the $SU(3)$ group, where $a = 1, 2, \ldots, 8$. It is understood that the repeated indices are summed over.

The last term in Eq. (2.4) governs the dynamics of the gauge sector and written in terms of the field strength tensor

$$F^a_{\mu\nu} = \partial_\mu A^a_\nu(x) - \partial_\nu A^a_\mu(x) - g_s f^{abc} A^b_\mu A^c_\nu, \tag{2.6}$$

where f^{abc} are the structure constants of the $SU(3)$ group. Last term survives because of the non-Abelian nature of the theory and it is responsible for the gluon-gluon interactions.

When quantizing the theory, two more terms corresponding to gauge fixing and ghost fields are added to Eq. (2.4) (see, for example, Ref. [6]). We are, however, omitting those terms in this prescription for the ease of discussion. In principle, another term which has the same dimension as the former terms and leaving the Lagrangian gauge invariant can be added but such a term breaks the CP-symmetry and becomes significant if one discusses the instanton field effects [7, 8], which we do not consider in this work.

QCD is now a well-established theory such that it is shown to be renormalisable, a crucial feature for a quantum field theory, by t'Hooft [9, 10] and it has a plausible property which clarifies the empirical facts arising from deep inelastic scattering experiments. In short, high energy $e-p$ scattering experiments revealed that within the proton there are point-like structures, which we now call quarks, and the scattering process can be calculated by considering a tree-level elastic $e-q$ interaction (see Chap. 9 of Ref. [11] for a detailed discussion). The physical implication is that these quarks act as free particles as if the strong interaction among themselves is non-existent. QCD (Yang-Mills theories in general) happens to have just the property, called the *Asymptotic Freedom* [12–15], providing a dynamical explanation to the weakness of the strong interactions in high-energy processes.

At this point, it is timely to shift our focus and discuss the renormalisation procedure briefly as it reveals the high and low energy nature of the QCD. Main idea behind the renormalisation of a quantum field theory is to regularise the ultraviolet divergences appearing when one considers the perturbative quantum corrections. Along the procedure one redefines the parameters of the theory, such as the bare coupling constant, bare mass and the bare fields, q and A_μ, to absorb the divergences. Redefinition of the parameters are actually done on some energy-scale conventionally denoted as μ. Fixing the value of μ depends on which energy scale we wish to investigate the theory therefore it introduces an arbitrariness but also allows to analyse the scale dependence of the theory or alternatively its parameters. Scale dependence of a parameter is probed by the method of renormalisation group founded by Gell-Mann and Low [16] in the context of QED but the idea is perfectly applicable to the QCD case. We will not, however, give a detailed account of the renormalisation procedure and the renormalisation group since we are only interested in illustrating the basic concept of the asymptotic freedom.

In order to see how the phenomenon of the asymptotic freedom appears we will look into the μ dependence of the renormalised strong coupling constant in the limit $\mu >> m_q$,

$$\mu\frac{d}{d\mu}g_s(\mu) = \beta(g_\mu) = -\beta_0 g_s^2(\mu) - \beta_1 g_s^3(\mu) - \cdots \tag{2.7}$$

where $\beta(g_\mu)$ is the Beta-function and this renormalisation group equation (RGE) is known as the Callan-Symanzik equation due to their work [17, 18]. Since we are interested in the coupling constant, Eq. (2.7) has only the coupling constant part. The total RGE has all the relevant terms related to the renormalised parameters that depend on the scale μ. The Beta-function is known up to 4-loops [19, 20] but we will restrict the discussion to lowest order which is enough for our purpose and omit the β_1 and higher-order terms. Equation (2.7) in the lowest order approximation is then given by

$$\mu\frac{d}{d\mu}g_s(\mu) = -\frac{g_s^2(\mu)}{4\pi}\frac{33-2n_f}{12\pi}, \tag{2.8}$$

and integrating with respect to μ we find,

$$\alpha_s(\mu) \equiv \frac{g_s^2(\mu)}{4\pi} = \frac{12\pi}{(33 - 2n_f)\ln\left(\mu^2/\Lambda_{QCD}^2\right)}, \tag{2.9}$$

where n_f is the number of active quark flavors at the chosen scale μ and Λ_{QCD} is an integration constant whose significance we will discuss below. There may be at most $n_f = 6$ different quark flavors at the practical energies so the overall sign of the $\alpha_s(\mu)$ is positive (an indication of the asymptotic freedom) and it is evident that the value of the strong coupling constant decreases as the energy (or the renormalisation) scale μ increases. This is the manifestation of the asymptotic freedom property of the QCD which, with its discovery, reinforced the belief that QCD is indeed the theory of strong interactions and due to the weakness of the interaction strength at higher energies, a perturbative approach is valid to investigate its physical implications.

In the discussion above, we have focused on the behaviour of the coupling constant in the high-energy region but the RGE holds information about the low-energy behaviour of the coupling as well. It is as we scale to the low-energy regime that the strong nature of QCD reveals itself. First thing to notice is that the Λ_{QCD} parameter in Eq. (2.9) marks the threshold at which scale the $\alpha_s(\mu)$ will diverge to infinity while along the way render the perturbative approach unusable since the strong interaction *will* become strong! α_s is found to exceed unity for $\mu \leq \mathcal{O}(0.1 - 1\text{GeV})$ [21]. Value of the Λ_{QCD}, on the other hand, lies in between $0.2 - 0.3\,\text{GeV}$ which is (from the grounds of dimensional analysis) close to $(200\,\text{MeV})^{-1} \sim 1\,\text{fm}$—the typical size of a hadron. In Fig. 2.1 we give the plot taken from the 2016 edition of the *Review of Particle Physics* [1] showing the behaviour of the strong coupling constant.

We have now made contact with the low-energy regime and are aware that the perturbative prescription breaks down. An important consequence of the low-energy dynamics of QCD is yet to reveal itself. Experimental evidence of the existence of quarks comes from the deep-inelastic scattering (DIS) experiments which can be accounted as an indirect probing method since the actual experiment involves a charged particle scattering from a composite object of quarks, anti-quarks and gluons, i.e. proton. The fact that no free (colored) quark has been observed and

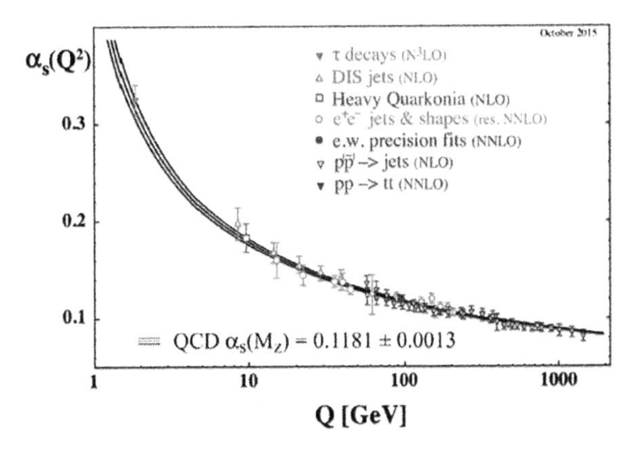

Fig. 2.1 Scale dependence of the strong coupling constant $\alpha_s(\mu \equiv Q^2)$ as well as various experimental determinations of its value. Curves are the perturbative-QCD predictions in the 4-loop approximation. We refer the reader to Ref. [1] and references therein

they are bound into colorless (color-singlet) composite objects suggests that the quarks (and anti-quarks) are confined into such objects, collectively called as hadrons. Confinement phenomenon is still a hypothesis waiting for a formal proof, however, our empirical and numerical understanding suggests that the color-color interaction is the reason. Phenomenologically the potential between a static quark-anti quark pair is parametrised by a Coulomb + linear-type potential reflecting the short-distance (high energy) and long-distance (low-energy) interactions. Experimental evidence for the Coulomb term has its origins from $p - \bar{p}$ collisions at CERN [22–24] and the linear term is deduced from hadron spectroscopy analysis (see Ref. [25] for a review). Numerical lattice calculations on the other hand provides a direct evidence of linear behaviour from first principles [25]. Naively the linear term implies that, strong-force lines are squeezed due to the self-interactions of the gluons and form tube-like structures, called flux tubes. These flux tubes are broken if one supplies the system with enough energy and a new quark-antiquark pair pops-up to form new flux tubes keeping the system as a color-singlet bound system. Although a delicate and interesting subject, we cut the confinement discussion short since rather than its origins or proof, we are interested in the outcome—hadrons.

2.4 Hadron Structure

Discovery of the composite nature of the hadrons dates back to the measurement of the magnetic moment of proton by Nobel laureate Stern in 1933 [26, 27]. The measured value of $\mu_p \sim 2.5\mu_N$ showed a significant deviation from unit nuclear magneton ($\mu_N = e/2M_N$), providing a clear hint of an inner structure. Later elastic *e*-*p* scattering experiments provided further evidence where, for example, the analyses in Ref. [28] indicate a clear discrepancy between the experimental data and theoretical cross-section calculations of a point-like particle. We show the cross-section plot taken from Ref. [28] in Fig. 2.2 where the deviation from point-like expectations is evident.

Pinpointing the existence of quarks is attributed to early deep-inelastic scattering experiments in SLAC, CERN SPS and Fermilab Tevatron with high energy and large momentum transfer (Q^2) between a lepton (electron, muon and neutrino) and a proton (see Ref. [29] and references therein). Higher momentum transfer means that the probe (a virtual photon in this case) has a finer resolution or a smaller wavelength $\lambda \sim 1/\sqrt{(}Q^2)$, which gives it the ability to probe smaller distances thus resolve the substructure of the proton. Cross-section formula for a DIS differs from the elastic case to incorporate the many hadron final states arising from the break up of the nucleon. Rather than simple Q^2-dependent form factors, it involves structure functions with two variables $W_{1,2}(Q^2, \nu)$ where the additional parameter $\nu = P \cdot q/M$ is the energy transferred to nucleon by the scattering electron. In the deep-inelastic energy limit, i.e. $Q^2 \to \infty$ and $\nu \to \infty$, however, it turns out that the structure functions scale as $\lim_{Q^2\to\infty} W_1(Q^2, \nu) = F_1(x)$ and $\lim_{Q^2\to\infty} \frac{\nu}{M} W_2(Q^2, \nu) = F_2(x)$ where the

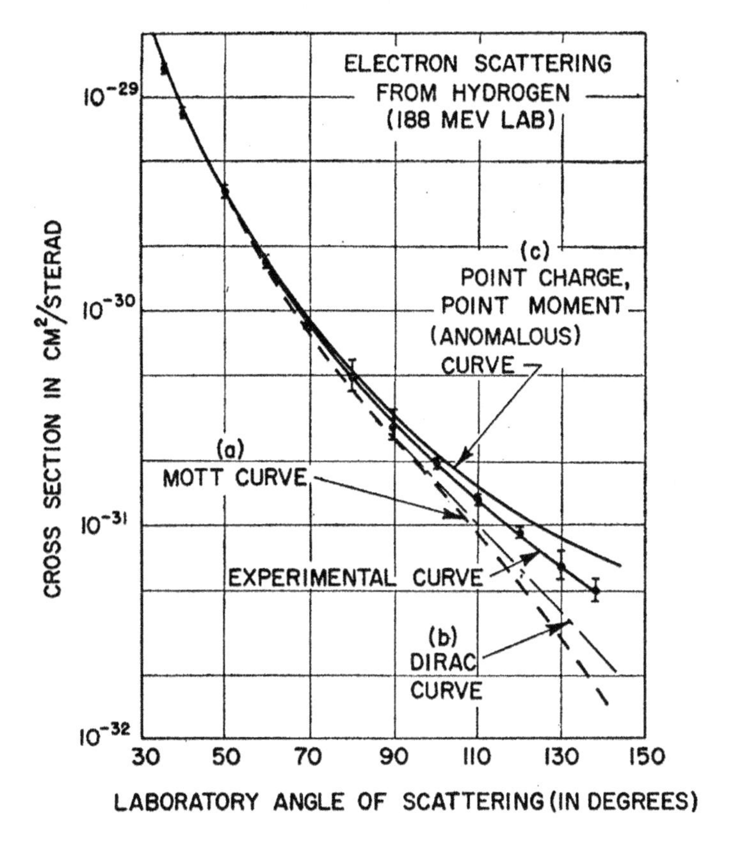

Fig. 2.2 Cross section of elastic *e-p* scattering plotted with respect to the scattering angle [28]. Theoretical curves **a**, **b** and **c** assumes a point-like proton, where **a** is the Mott curve and **b** and **c** uses Rosenbluth formalism. Experimental curve is a best fit to Rosenbluth formula with the form factor interpretation

Bjorken scaling parameter is defined as $x = Q^2/2M\nu$ [30]. Weak Q^2-dependence observed in early SLAC DIS data was in accordance with the Bjorken scaling expectations and has led to the development of the parton model by Feynman [31, 32] which suggests elastic scattering of virtual photon from free point-like constituents carrying some fraction of the total energy of the proton. Parton model calculations combined with experiments pioneered the direct identification of the quarks which were predicted earlier by Gell-Mann [33] and Zweig [34]. Data showed that nucleon is indeed composed of three valence quarks but it has also revealed that there is actually a *sea* of quarks, anti-quarks and gluons changing the simple picture into a highly dynamical one.

It is instructive at this point to summarise how one forms a field theoretical framework to understand the composite nature of the hadrons by studying the elastic *e-p* scattering. If proton was indeed a point particle, we would describe the interaction cross-sections by the Mott formula,

$$\left(\frac{d\sigma}{d\Omega}\right)_{Mott} = \frac{(Z\alpha)^2E^2}{4k^2\sin^4(\theta/2)}\left(1-\frac{k^2}{E^2}\sin^2(\theta/2)\right), \tag{2.10}$$

where Z is the number of protons of the fixed target, α is the fine-structure constant of QED and E and k are the energy and momentum of the incoming electron. θ stands for the scattering angle related to transferred momentum q^2 and out-going electron energy as $q^2 = -4EE'\sin^2(\theta/2)$. However, it is clear that the point-like assumption does not hold so we should include a term to incorporate the inner structure by multiplying the Mott formula with a *form factor* term,

$$\left(\frac{d\sigma}{d\Omega}\right) = \left(\frac{d\sigma}{d\Omega}\right)_{Mott} \left|F(q^2)\right|^2. \tag{2.11}$$

We have to estimate the form of the $F(q^2)$ term however, in order to make reliable calculations. Since this is an electromagnetic interaction, we will use the perturbative QED formalism and consider a tree-level one-photon exchange interaction. Note that, two-photon exchange processes also have a significant role [35] and should be included for rigorous analyses, however, we will not consider those higher-order interactions since a tree-level approximation is adequate to make our point. We write down the S-matrix of the tree-level interaction shown in Fig. 2.3 as,

$$\begin{aligned} S &= (2\pi)^4\delta^4(k+P-P'-k')\bar{u}(k')(-ie\gamma^\mu)u(k)\frac{-i}{q^2}\langle P'|(ie)J^\mu|P\rangle \\ &= -i(2\pi)^4\delta^4(k+P-P'-k')\mathcal{M}, \end{aligned} \tag{2.12}$$

where the Dirac-delta functions ensure the energy-momentum conservation, $\bar{u}(k')$ and $u(k)$ are the fermion spinor fields with four-momentum k and k', $(-ie\gamma^\mu)$ and $(-iq^2)$ are the vertex factors of the electron and proton vertices and the last term in brakets is the hadronic matrix element with the electromagnetic current inserted in

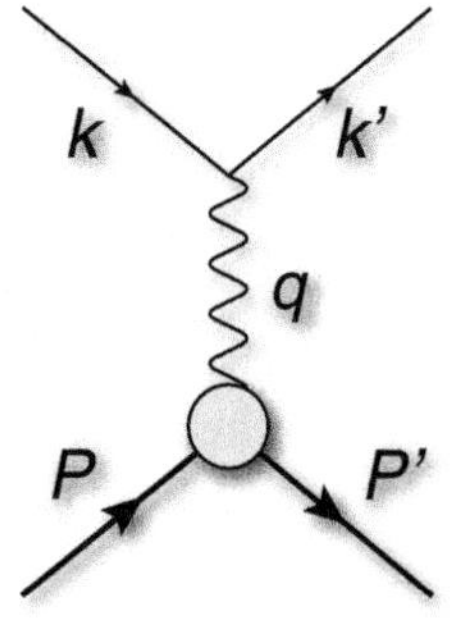

Fig. 2.3 Tree-level Feynman diagram of an elastic *e-p* scattering. Blob indicates the unknown structure of the proton

between the states. We have introduced the invariant amplitude,

$$\mathcal{M} = \frac{1}{q^2}\bar{u}(k')(-ie\gamma^{\mu})u(k)\langle P'|(ie)J^{\mu}|P\rangle, \tag{2.13}$$

on the second line for the ease of further discussion and the external electromagnetic current is given by,

$$J^{\mu} = \sum_i e_i \bar{\psi}_i \gamma^{\mu} \psi_i, \tag{2.14}$$

where the index i sums over all valence- and sea-quark flavours that are within the proton. Interaction cross-section in terms of the invariant amplitude $\mathcal{M}$ follows as,

$$\frac{d\sigma}{d\Omega} = \frac{E'}{8\pi^2 EM^2} \frac{|\mathcal{M}|^2}{1 + \frac{2E}{M}\sin^2(\theta/2)}, \tag{2.15}$$

where the squared invariant amplitude is written in terms of leptonic and hadronic tensors as,

$$|\mathcal{M}|^2 = \frac{e^4}{Q^4}\left[\bar{u}(k')\gamma^{\mu}u(k)\right]\left[\bar{u}(k)\gamma^{\nu}u(k')\right]^* \langle P|J^{\nu}|P'\rangle\langle P'|J^{\mu}|P\rangle, \tag{2.16}$$

with $-q^2 = Q^2$. Leptonic part can be calculated by perturbative means since it is a QED process but since we do not know its exact form we have to parametrise the hadronic piece while keeping the Lorentz-invariance intact. Focusing on one of the matrix elements, we may form independent Lorentz-invariant structures using the elements we have at hand, namely the four-vectors P^{μ}, P'^{μ}, q^{μ} and Dirac γ-matrices (excluding γ_5 due to parity conservation), and write down a widely used parametrisation,

$$\langle P'|J^{\mu}(\mathbf{q})|P\rangle = \bar{u}(P')\left[\gamma^{\mu}F_1(q^2) + i\sigma^{\mu\nu}\frac{q_{\nu}}{2M}F_2(q^2)\right]u(P), \tag{2.17}$$

with $\sigma^{\mu\nu} = \frac{1}{2}\{\gamma^{\mu}, \gamma^{\nu}\}$ and M being the mass of the nucleon. F_1 and F_2 are known as Dirac and Pauli form factors respectively and are associated with the charge and anomalous magnetic moment of the nucleon. Linear combinations of Dirac and Pauli form factors are used to define the Sachs electric and magnetic form factors,

$$\begin{aligned} G_E(Q^2) &= F_1(Q^2) - \tau F_2(Q^2), \\ G_M(Q^2) &= F_1(Q^2) + F_2(Q^2), \end{aligned} \tag{2.18}$$

where $\tau = Q^2/4M^2$ and in terms of these form factors the elastic scattering cross-section in the lab frame is given by,

$$\frac{d\sigma}{d\Omega} = \sigma_{\text{Mott}} \left[\frac{G_E^2(Q^2) + \tau G_M^2(Q^2)}{1+\tau} + 2\tau G_M^2(Q^2)\tan^2\frac{\theta}{2} \right]. \tag{2.19}$$

Rewriting the cross-section in terms of the virtual photon's longitudinal polarization $\epsilon = (1 + (1+\tau)2\tan^2(\theta/2))^{-1}$, we end up with the Rosenbluth formula [36],

$$\frac{d\sigma}{d\Omega} = \frac{\sigma_{\text{Mott}}}{1+\tau} \left[G_E^2(Q^2) + \frac{\tau}{\epsilon} G_M^2(Q^2) \right]. \tag{2.20}$$

Experimental data can be analysed by the above formula such that the information about electric and magnetic form factors can be accessed respectively from the slope and intersect of a curve fitted to cross-section versus scattering angle data at a fixed momentum transfer Q^2. Such an analysis is known as the Rosenbluth seperation technique. There are many caveats and technicalities to mention from an experimental point-of-view and the above account is by no means complete however very useful in making contact with a crucial theoretical object which sits in the center of the calculations that we are interested in for the rest of this work—the matrix element.

In Eq. (2.17) we have written down the parametrisation of the hadronic matrix element (ME) for an electromagnetic spin$-1/2 \rightarrow$ spin$-1/2$ transition processes in a general sense. Given that the MEs hold information about the hadron, we are already in the non-perturbative regime of QCD so that perturbative calculations are unreliable, however, it is possible to calculate the MEs by non-perturbative approaches. The most promising non-perturbative method we have at hand is lattice QCD which we utilise to calculate the MEs and extract the form factors throughout this work. We will summarise the lattice formulation of QCD in the next Chapter but, before that, let us close this section by mentioning the physical meaning of the form factors.

In analogy to non-relativistic physics, a form factor can be considered as a three-dimensional Fourier transform of the density distribution of a quantity of a hadron under certain conditions. Considering a small momentum transfer, Q^2, between the baryon and the external current or a limit where the mass of the baryon is much larger than the transferred momentum, $Q^2 \ll M^2$, we have the assumption that the internal structure of the initial and final states remain the same. However, with increasing Q^2, recoil effects become significant so that the wave functions of the initial and final states differ, rendering a density distribution interpretation questionable. In order to have a rigorous definition, one considers the Breit frame where the magnitude of the initial and final momentum of the hadron is equal, i.e. $P' = P$. Breit frame configuration helps us to recover the density-distribution interpretation of the form factors. For an electromagnetic case as above, the Sachs form factors G_E and G_M are related to the charge and magnetisation densities of the hadron [37] in the Breit frame where for example the G_E is given by the Fourier transformation of an electric charge density as,

$$G_E(Q^2) = \int d^3\mathbf{x} e^{i\mathbf{x}\mathbf{q}} \rho(\mathbf{x}) \simeq G_E(0) \left(1 - \frac{1}{6} Q^2 \langle r_E^2 \rangle + \dots \right), \tag{2.21}$$

where the first term in parenthesis reduces to the total electric charge of the hadron and the second term is the definition of the square of the electric root-mean square radius. We sketch further details in Sect. 5.2.2 and interpretations of the quantities extracted from form factors are discussed in Sects. 5.3 and 5.4.

References

1. C. Patrignani et al., Review of particle physics. Chin. Phys. **C40**(10), 100001 (2016), https://doi.org/10.1088/1674-1137/40/10/100001
2. O.W. Greenberg, Spin and unitary-spin independence in a paraquark model of baryons and mesons. Phys. Rev. Lett. **13**, 598–602 (1964), http://link.aps.org/doi/10.1103/PhysRevLett.13.598
3. Y. Nambu, A systematics of hadrons in subnuclear physics, in *Preludes in Theoretical Physics in Honor of V. F. Weisskopf* (1966), p. 133
4. M.Y. Han, Y. Nambu, Three-triplet model with double SU(3) symmetry. Phys. Rev. **139**, B1006–B1010 (1965), http://link.aps.org/doi/10.1103/PhysRev.139.B1006
5. L. Montanet et al., Review of particle properties. Phys. Rev. D **50**, 1173–1814 (1994), http://link.aps.org/doi/10.1103/PhysRevD.50.1173
6. L.D. Faddeev, V.N. Popov, Feynman diagrams for the yang-mills field. Phys. Lett. B **25**(1), 29–30 (1967). ISSN 0370-2693, http://www.sciencedirect.com/science/article/pii/0370269367900676
7. T.-P. Cheng, L.-F. Li, *Gauge Theory of Elementary Particle Physics* (Clarendon Press, Oxford, 1984)
8. S. Weinberg, *The Quantum Theory of Fields. Vol. 2: Modern Applications* (Cambridge University Press, Cambridge, 2013). ISBN 9781139632478, 9780521670548, 9780521550024
9. G. Hooft, Renormalization of massless yang-mills fields. Nucl. Phys. B **33**(1), 173–199 (1971a)
10. G. Hooft, Renormalizable lagrangians for massive yang-mills fields. Nucl. Phys. B **35**(1), 167–188 (1971b)
11. I.J.R. Aitchison, AJG Hey, *Gauge Theories in Particle Physics: A Practical Introduction*, 4th edn. (CRC Press, Boca Raton, FL, 2013), https://cds.cern.ch/record/1507184
12. J. Frenkel, J.C. Taylor, Asymptotic freedom in the axial and coulomb gauges. Nucl. Phys. B **109**(3), 439–451 (1976). ISSN 0550-3213, http://www.sciencedirect.com/science/article/pii/0550321376902443
13. I.B. Khriplovich, Green's functions in theories with non-abelian gauge group. Sov. J. Nucl. Phys. **10**, 235–242 (1969). Yad. Fiz.10,409 (1969)
14. D.J. Gross, F. Wilczek, Ultraviolet behavior of non-abelian gauge theories. Phys. Rev. Lett. **30**, 1343–1346 (1973), http://link.aps.org/doi/10.1103/PhysRevLett.30.1343
15. H.D. Politzer, Reliable perturbative results for strong interactions? Phys. Rev. Lett. **30**, 1346–1349 (1973), http://link.aps.org/doi/10.1103/PhysRevLett.30.1346
16. M. Gell-Mann, F.E. Low, Quantum electrodynamics at small distances. Phys. Rev. **95**, 1300–1312 (1954), http://link.aps.org/doi/10.1103/PhysRev.95.1300
17. C.G. Callan, Broken scale invariance in scalar field theory. Phys. Rev. D **2**, 1541–1547 (1970), http://link.aps.org/doi/10.1103/PhysRevD.2.1541
18. K. Symanzik, Small distance behaviour in field theory and power counting. Comm. Math. Phys. **18**(3), 227–246 (1970), http://projecteuclid.org/euclid.cmp/1103842537
19. M. Czakon, The four-loop QCD beta-function and anomalous dimensions. Nucl. Phys. B **710**, 485–498 (2005), https://doi.org/10.1016/j.nuclphysb.2005.01.012
20. T. van Ritbergen, J.A.M. Vermaseren, S.A. Larin, The four loop beta function in quantum chromodynamics. Phys. Lett. B **400**, 379–384 (1997), https://doi.org/10.1016/S0370-2693(97)00370-5

21. S. Bethke, The 2009 world average of α s. Eur. Phys. J. C **64**(4), 689–703 (2009). ISSN 1434-6052, http://dx.doi.org/10.1140/epjc/s10052-009-1173-1
22. G. Arnison et al., Observation of jets in high transverse energy events at the cern proton antiproton collider. Phys. Lett. B **123**(1), 115–122 (1983). ISSN 0370-2693, http://www.sciencedirect.com/science/article/pii/037026938390970X
23. G. Arnison et al., Associated production of an isolated, large-transverse-momentum lepton (electron or muon), and two jets at the cern pp collider. Phys. Lett. B **147**(6), 493–508 (1984). ISSN 0370-2693, http://www.sciencedirect.com/science/article/pii/0370269384914102
24. P. Bagnaia et al., Measurement of very large transverse momentum jet production at the cern pp collider. Phys. Lett. B **138**(5), 430–440 (1984). ISSN 0370-2693, http://www.sciencedirect.com/science/article/pii/037026938491935X
25. G.S. Bali, QCD forces and heavy quark bound states. Phys. Rept. **343**, 1–136 (2001), https://doi.org/10.1016/S0370-1573(00)00079-X
26. I. Estermann, O. Stern, Über die magnetische ablenkung von wasserstoffmolekülen und das magnetische moment des protons. ii. Zeitschrift für Physik **85**(1), 17–24 (1933). ISSN 0044-3328, http://dx.doi.org/10.1007/BF01330774
27. R. Frisch, O. Stern, Über die magnetische ablenkung von wasserstoffmolekülen und das magnetische moment des protons. i. Zeitschrift für Physik **85**(1), 4–16 (1933). ISSN 0044-3328, http://dx.doi.org/10.1007/BF01330773
28. R.W. McAllister, R. Hofstadter, Elastic scattering of 188-mev electrons from the proton and the alpha particle. Phys. Rev. **102**, 851–856 (1956), http://link.aps.org/doi/10.1103/PhysRev.102.851
29. E.M. Riordan, The Discovery of quarks. Science **256**, 1287–1293 (1992), https://doi.org/10.1126/science.256.5061.1287
30. J.D. Bjorken, Asymptotic sum rules at infinite momentum. Phys. Rev. **179**, 1547–1553 (1969), http://link.aps.org/doi/10.1103/PhysRev.179.1547
31. R.P. Feynman, *Photon-Hadron Interactions* (WA Benjamin Inc., Reading, MA, 1972)
32. R.P. Feynman, Very high-energy collisions of hadrons. Phys. Rev. Lett. **23**, 1415–1417 (1969), http://link.aps.org/doi/10.1103/PhysRevLett.23.1415
33. M. Gell-Mann, A schematic model of baryons and mesons. Phys. Lett. **8**(3), 214–215 (1964). ISSN 0031-9163, http://www.sciencedirect.com/science/article/pii/S0031916364920013
34. G. Zweig, An SU(3) model for strong interaction symmetry and its breaking. Version 2, in *Developments in the quark theory of hadrons, 1964–1978*, vol. 1, ed. by D.B. Lichtenberg, S.P. Rosen (Hadronic Press, Inc., Nonantum, MA, 1980), pp. 22–101, http://inspirehep.net/record/4674/files/cern-th-412.pdf
35. J. Arrington, P.G. Blunden, W. Melnitchouk, Review of two-photon exchange in electron scattering. Prog. Part. Nucl. Phys. **66**(4), 782–833 (2011). ISSN 0146-6410, http://www.sciencedirect.com/science/article/pii/S0146641011000962
36. M.N. Rosenbluth, High energy elastic scattering of electrons on protons. Phys. Rev. **79**, 615–619 (1950), http://link.aps.org/doi/10.1103/PhysRev.79.615
37. R.G. Sachs, High-energy behavior of nucleon electromagnetic form factors. Phys. Rev. **126**, 2256–2260 (1962), http://link.aps.org/doi/10.1103/PhysRev.126.2256

Chapter 3
Lattice Formulation of QCD

Abstract In this chapter, we provide a detailed formulation of the lattice field theory in the context of QCD. Basic changes compared to the continuum formulation is introduced. Then, we formulate the Euclidean QCD action starting from a naive approach and improve it step-by-step until we have a suitable lattice action. We discuss the gauge and fermion sectors individually with their respective challenges and improvements. Steps of a typical application of the method are outlined in the closing of the chapter.

Keywords Lattice field theory · Euclidean space-time · Discrete action
Fermion doubling · Improvement program

3.1 Euclidean Action

Quantisation of QCD is made via the Feynman's path integral approach [1]. Formally, expectation value of a physical observable is given by a functional integral,

$$\langle \hat{\mathcal{O}} \rangle = \frac{1}{Z} \int \mathcal{D}\left[\bar{\psi}, \psi, A\right] e^{i S_{QCD}[\bar{\psi},\psi,A]} \mathcal{O}\left[\bar{\psi}, \psi, A\right], \tag{3.1}$$

where the LHS is in operator language and the RHS contains the classical action and fields. QCD action in the Minkowski space is given by,

$$S^{M}_{QCD}[\psi, \bar{\psi}, A] = \int d^4x \left\{ \sum_q \bar{\psi}_q \left(i\mathcal{D}_\mu \gamma^\mu - m_q \right) \psi_q - \frac{1}{4} F^a_{\mu\nu} F^{\mu\nu}_a \right\}, \tag{3.2}$$

and the partition function is defined as,

$$Z = \int \mathcal{D}\left[\bar{\psi}, \psi, A\right] e^{i S_{QCD}[\bar{\psi},\psi,A]}. \tag{3.3}$$

K. U. Can, *Electromagnetic Form Factors of Charmed Baryons in Lattice QCD*,
Springer Theses, https://doi.org/10.1007/978-981-10-8995-4_3

In the lattice approach, one evaluates the integral in Eq. (3.1) numerically by means of Monte Carlo methods. However, the sampling weight, $e^{iS_{QCD}[\bar{\psi},\psi,A]}$, is an imaginary function with a highly oscillatory behaviour which renders a reliable numerical treatment rather challenging. In order to tame the oscillating function, we perform a Wick rotation, i.e. we rotate from a Minkowski-space to Euclidean one by the, $t \rightarrow -it$, transformation and the action transforms as $iS \rightarrow -S_E$. This rotation basically allows us to connect the lattice theory with statistical mechanics, which we briefly discuss in Sect. 3.5. Corresponding Euclidean QCD action then takes the form,

$$iS^M_{QCD} \rightarrow -S^E_{QCD}[\psi, \bar{\psi}, A] = -\int d^4x \left\{ \sum_q \bar{\psi}_q \left(\mathcal{D}_\mu \gamma^\mu + m_q \right) \psi_q + \frac{1}{4} \left(F^a_{\mu\nu} \right)^2 \right\}. \tag{3.4}$$

One, of course, has to define the discrete form of the action for a numerical approach. In the following sections we discretise the space-time continuum and derive the discrete QCD action.

A final remark is on the Dirac index convention. It changes from $\mu = 0, 1, 2, 3 \equiv (t, x, y, z)$ to $\mu = 1, 2, 3, 4 \equiv (x, y, z, t)$ and we use a chiral basis for the gamma matrices,

$$\gamma_1 = -i\gamma_1^D, \quad \gamma_2 = -i\gamma_2^D, \quad \gamma_3 = -i\gamma_3^D, \quad \gamma_4 = \gamma_0^D, \tag{3.5}$$

which satisfy, $\{\gamma_\mu, \gamma_\nu\} = 2\delta_{\mu\nu}$. The usual Minkowski space Dirac gamma matrices are,

$$\begin{aligned}
\gamma_0^D &= \begin{pmatrix} 1 & 0 & 0 & 0 \\ 0 & 1 & 0 & 0 \\ 0 & 0 & -1 & 0 \\ 0 & 0 & 0 & -1 \end{pmatrix} & \gamma_1^D &= \begin{pmatrix} 0 & 0 & 0 & 1 \\ 0 & 0 & 1 & 0 \\ 0 & -1 & 0 & 0 \\ -1 & 0 & 0 & 0 \end{pmatrix} \\
\gamma_2^D &= \begin{pmatrix} 0 & 0 & 0 & -i \\ 0 & 0 & i & 0 \\ 0 & i & 0 & 0 \\ -i & 0 & 0 & 0 \end{pmatrix} & \gamma_3^D &= \begin{pmatrix} 0 & 0 & 1 & 0 \\ 0 & 0 & 0 & -1 \\ -1 & 0 & 0 & 0 \\ 0 & 1 & 0 & 0 \end{pmatrix}.
\end{aligned} \tag{3.6}$$

3.2 Discrete Space-Time

There are two main elements of the theory to discretise in the first place: Space-time continuum itself and the fermion and gauge fields. We start by replacing the continuous space-time by a 4D discrete lattice. A straightforward discretisation is to set the lattice spacing equal on all dimensions, i.e. $a = a_S = a_T$, for a lattice of size $N_S \times N_S \times N_S \times N_T$. Any point on the discrete space-time is then given by,

$$n = (n_1, n_2, n_3, n_4), \quad | \quad n_{1,2,3} = 0, 1, \ldots, N_S - 1, \quad n_4 = 0, 1, \ldots, N_T - 1, \tag{3.7}$$

where N_S and N_T are the number of spatial and temporal steps, respectively. Depending on the application or available resources, spatial and temporal lattice spacing and/or extent may be taken different, e.g. $a_S \neq a_T$ and/or $N_T > N_S$. For example modern-day spectroscopy calculations usually prefer longer temporal extent, sometimes with $a_S \neq a_T$, to isolate a clear signal whereas $N_T < N_S$ lattices are employed in finite temperature studies.

Fermion fields are restricted to live on lattice sites, n, and allowed to move step by step on straight lines only,

$$\psi(x) \to a^{-3/2}\psi(an) \qquad \bar{\psi}(x) \to a^{-3/2}\bar{\psi}(an), \tag{3.8}$$

$$\psi(an') = \psi(an \pm a\hat{\mu}) \qquad \bar{\psi}(an') = \bar{\psi}(an \pm a\hat{\mu}) \tag{3.9}$$

where x corresponds to the continuum coordinate and $\hat{\mu}$ denotes the unit vector in the μ direction. We will drop the a from now on for simplicity.

We replace the gauge fields by *Link Variables*,

$$U(n, n \pm \hat{\mu}) \equiv U_{\pm\mu}(n), \quad U_{-\mu}(n) \equiv U^{\dagger}_{\mu}(n - \hat{\mu}), \tag{3.10}$$

which assume the role of connecting adjacent lattice sites to each other. They are defined as 3×3 $SU(3)_{\text{color}}$ matrices,

$$U_\mu = \begin{matrix} & \begin{matrix} r & g & b \end{matrix} \\ \begin{matrix} r \\ g \\ b \end{matrix} & \begin{pmatrix} U^{rr}_\mu & U^{rg}_\mu & U^{rb}_\mu \\ U^{gr}_\mu & U^{gg}_\mu & U^{gb}_\mu \\ U^{br}_\mu & U^{bg}_\mu & U^{bb}_\mu \end{pmatrix} \end{matrix}, \tag{3.11}$$

where each element of the matrix corresponds to probability density of a transition from one color component to another. Link variables are related to the continuum gauge fields by an exponent,

$$U_\mu(n) = \exp\left(iaA_\mu(n)\right). \tag{3.12}$$

Since we are defining a finite system, boundaries come into consideration. We can impose either periodic, anti-periodic or fixed-boundary—Dirichlet or Neumann—boundary conditions depending on the application. For most of the applications, including ours, (anti-)periodic boundaries suffice while the Schrödinger functional method [2–6] can be counted as an example to the use of Dirichlet boundary conditions. Usually periodic boundaries are imposed on all directions for the gauge fields while one direction, generally the temporal direction, is chosen to be anti-periodic for fermions to account for Fermi-Dirac statistics. We can write the (anti-)periodic boundary conditions (for a single dimension) in mathematical notation as,

$$\begin{aligned}
\psi(0,n_2,n_3,n_4) &= \psi(N_S,n_2,n_3,n_4) & U_\mu(N_S,n_2,n_3,n_4) &= U_\mu(0,n_2,n_3,n_4),\\
\psi(n_1,0,n_3,n_4) &= \psi(n_1,N_S,n_3,n_4) & U_\mu(n_1,N_S,n_3,n_4) &= U_\mu(n_1,0,n_3,n_4),\\
\psi(n_1,n_2,0,n_4) &= \psi(n_1,n_2,N_S,n_4) & U_\mu(n_1,n_2,N_S,n_4) &= U_\mu(n_1,n_2,0,n_4),\\
\psi(n_1,n_2,n_3,0) &= -\psi(n_1,n_2,n_3,N_T) & U_\mu(n_1,n_2,n_3,N_T) &= U_\mu(n_1,n_2,n_3,0).
\end{aligned}$$

An important implication of the discrete space-time is the quantisation of the momentum on the lattice. The fact that the fields are restricted to lattice sites and can only move in discrete steps, imposes the condition,

$$p = \frac{2\pi n}{N_S a}, \quad n = 1, \ldots, N_S \tag{3.13}$$

on the lattice momentum. We immediately see that the largest allowed momenta is $p = 2\pi/a$ which implies that the lattice spacing acts an inverse UV cutoff, providing a UV-regularisation of a field theory formulated on a lattice.

3.3 Gauge Action

3.3.1 Naive Discretisation

We want to form the simplest discrete gauge action that reduces to the continuum gauge action in the $a \to 0$ limit while maintaining the local $SU(3)$ gauge invariance of QCD. Gauge part is composed of gauge fields only and the corresponding lattice objects we have at hand are link variables, $U_\mu(n)$. Under a local gauge transformation $V(n)$, link variables transform as,

$$U_\mu(n) \to U'_\mu(n) = V(n)U_\mu(n)V^\dagger(n+\hat{\mu}), \tag{3.14}$$

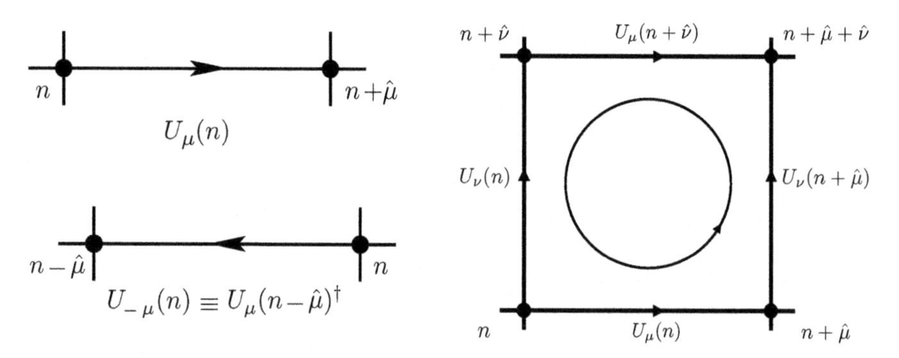

Fig. 3.1 Forward and backward link variables (left) and 1×1 Plaquette $U_{\mu\nu}$, constructed from four link variables (right)

which has the same geometrical interpretation of the continuum counterpart. Then the simplest gauge-invariant object that we can define is a 1×1 loop of link variables,

$$U_{\mu\nu}(n) = U_\mu(n)U_\nu(n+\hat{\mu})U^\dagger_\mu(n+\hat{\nu})U^\dagger_\nu(n), \tag{3.15}$$

known as a plaquette. Forward and backward link variables and the plaquette object are illustrated in Fig. 3.1.

In order to reveal the connection between the plaquette and the continuum field strength tensor, we replace the link variables with Eq. (3.12),

$$\begin{aligned} U_{\mu\nu}(n) &= U_\mu(n)U_\nu(n+\hat{\mu})U^\dagger_\mu(n+\hat{\nu})U^\dagger_\nu(n) \\ &= e^{iag_sA_\mu(n)}e^{iag_sA_\nu(n+\hat{\mu})}e^{-iag_sA_\mu(n+\hat{\nu})}e^{-iag_sA_\nu(n)} \\ &= \exp\Big\{iag_sA_\mu(n) + iag_sA_\nu(n+\hat{\mu}) - iag_sA_\mu(n+\hat{\nu}) - iag_sA_\nu(n) \\ &\quad - \frac{a^2g_s^2}{2}[A_\mu(n), A_\nu(n+\hat{\mu})] - \frac{a^2g_s^2}{2}[A_\mu(n+\hat{\nu}), A_\nu(n)] \\ &\quad + \frac{a^2g_s^2}{2}[A_\nu(n+\hat{\mu}), A_\mu(n+\hat{\nu})] + \frac{a^2g_s^2}{2}[A_\mu(n), A_\nu(n)] \\ &\quad + \frac{a^2g_s^2}{2}[A_\mu(n), A_\mu(n+\hat{\nu})] + \frac{a^2g_s^2}{2}[A_\nu(n+\hat{\mu}), A_\nu(n)] + \mathcal{O}(a^3)\Big\} \end{aligned} \tag{3.16}$$

where we have used the Baker-Campbell-Hausdorff formula,

$$\exp(A)\exp(B) = \exp\left(A + B + \frac{1}{2}[A, B] + \dots\right), \tag{3.17}$$

to work out the exponentials and included terms up to $\mathcal{O}(a^2)$ only. Now by replacing the displaced gauge fields by their Taylor expansions,

$$A_\nu(n+\hat{\mu}) = A_\mu(n) + a\partial_\mu A_\nu(n) + \mathcal{O}(a^2), \tag{3.18}$$

we obtain,

$$\begin{aligned} U_{\mu\nu} &= \exp\left\{ia^2g_s\left(\partial_\mu A_\nu(n) - \partial_\nu A_\mu(n) + ig_s\left[A_\mu(n), A_\nu(n)\right]\right) + \mathcal{O}(a^3)\right\} \\ &= \exp\left\{ia^2g_sF_{\mu\nu}(n) + \mathcal{O}(a^3)\right\} \\ &= 1 + ia^2g_sF_{\mu\nu}(n) - \frac{a^4g_s^2}{2}(F_{\mu\nu}(n))^2 + \mathcal{O}(a^6). \end{aligned} \tag{3.19}$$

Real part of the plaquette isolates the term that we are interested in,

$$\mathrm{ReTr}\left(1 - U_{\mu\nu}\right) = \frac{a^4g_s^2}{2}(F_{\mu\nu}(n))^2 + \text{higher orders in } a, \tag{3.20}$$

where the trace acts on color indices and ensures the gauge invariance of the expression. Finally, summing over all lattice sites and plaquettes associated with each site, and including the color factor, we define the Wilson gauge action as,

$$S_G[U] = \frac{\beta}{3} \sum_n \sum_{\mu<\nu} \mathrm{ReTr}\left(1 - U_{\mu\nu}(n)\right) = a^4 \sum_n \sum_{\mu<\nu} \frac{1}{2} \mathrm{Tr}\left[F_{\mu\nu}(n)^2\right] + \mathcal{O}(a^2), \tag{3.21}$$

$$\rightarrow \int d^4x \sum_{\mu<\nu} \frac{1}{2} \mathrm{Tr}\left[F_{\mu\nu}(n)^2\right] + \mathcal{O}(a^2). \tag{3.22}$$

where the *inverse coupling* is given by $\beta = 2N_c/g_s^2$.

Naive discretisation matches the continuum gauge action at $\mathcal{O}(a^2)$ level, or in other words, has $\mathcal{O}(a^2)$ discretisation errors. Since the action is not unique, we can add higher dimensional terms to improve on the errors, or other properties, as long as we recover the continuum action in the $a \rightarrow 0$ limit.

3.3.2 Improved Action

We will summarise the improved gauge action used by the PACS-CS Collaboration to generate the gauge configurations [7], relevant for this work. CP-PACS Collaboration investigates the improved actions and they note that the Iwasaki gauge action leads to a better rotational symmetry, beneficial for static quark potential studies, and reduces the coefficients of the $\mathcal{O}(a^2)$ errors [8].

In addition to the Wilson plaquette, we can form dimension-6 gauge-invariant objects composed of six link variables as shown in Fig. 3.2. In principle, dimension-5 objects are the next in terms of dimension counting, however those object, formed only by link variables, are not gauge-invariant. Gauge action with the addition of dimension-6 operators is written as,

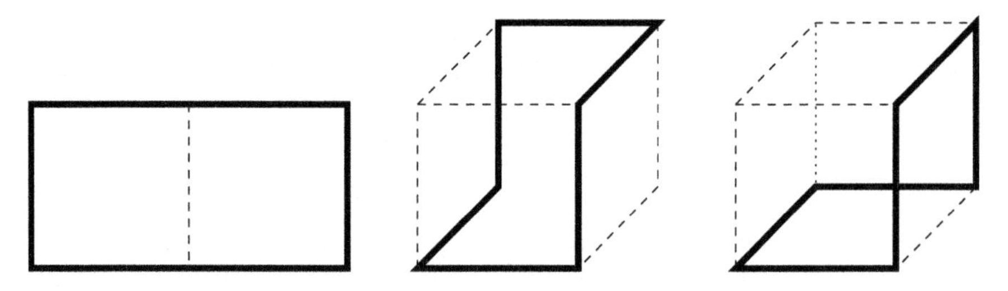

Fig. 3.2 Dimension-6 rectangular (1×2), twisted and L-shaped Wilson loops

$$S_G[U] = \frac{\beta}{6}\left(c_0 \sum_{n,\mu<\nu} W^{(4)}_{\mu\nu}(n) + \sum_i^3 c_i \sum_{n,\mu<\nu} W^{(6)}_{\mu\nu}(n)\right), \tag{3.23}$$

where the first term is the dimension-4 Wilson loop (plaquette) and the summation over i includes the dimension-6 rectangular, twisted and L-shaped Wilson loops, respectively. Iwasaki includes the rectangular term only in his approach and uses an approximate block-spin renormalisation group analysis of Wilson loops to fix its coefficient $c_1 = -0.331$. Coefficient of the plaquette term follows from the normalisation condition, $c_0 = 1 - 8c_1 = 3.648$, so the Iwasaki action takes the form [9]:

$$S_G[U] = \frac{\beta}{6}\left(c_0 \sum_{n,\mu<\nu} W^{1\times1}_{\mu\nu}(n) + c_1 \sum_{n,\mu<\nu} W^{1\times2}_{\mu\nu}(n)\right). \tag{3.24}$$

Lüscher and Weisz show that in tree-level approximation coefficients of twisted and L-shaped contributions can be taken $c_2 = c_3 = 0$. They study 1-loop corrections and introduce the 1-loop improved action with $\mathcal{O}(g^4a^4)$ corrections which has the coefficients $c_0(g^2) = 5/3 + 0.2370g^2$, $c_1(g^2) = -1/12 - 0.02521g^2$, $c_2(g^2) = -0.00441g^2$ and $c_3(g^2) = 0$ [10]. Small c_3 coefficient suggests that improving the action by adding only the rectangular terms already provides a good approximation of the continuum action.

3.4 Fermion Action

Continuing on with the fermionic part of the QCD action, we introduce the Wilson's fermion action and its improved version, the Sheikholeslami-Wohlert (Clover) action which we employ in this work.

3.4.1 Naive Discretisation

First, define the discrete partial derivative by the central-difference limiting procedure,

$$\partial_\mu \psi(n) = \frac{\psi(n+\hat{\mu}) - \psi(n-\hat{\mu})}{2a}, \tag{3.25}$$

where $\hat{\mu}$ is the unit vector in the μ direction and a is the lattice spacing. Applying the above formula to the Dirac equation for a single quark flavor, we have,

$$S_F[\psi, \bar{\psi}] = a^4 \sum_n \bar{\psi}(n) \left[\sum_{\mu=1}^{4} \gamma_\mu \frac{\psi(n+\hat{\mu}) - \psi(n-\hat{\mu})}{2a} + m_q \psi(n)\right], \tag{3.26}$$

where four-dimensional integral is replaced by a sum over all lattice points, $a^4 \sum_{\mathbf{n}}$, as in the gauge action case. This description however is not gauge-invariant under local $SU(3)_c$ transformations since fermion fields, $\psi(n+\hat{\mu})$ and $\psi(n-\hat{\mu})$, acquire different phases under transformations defined by,

$$\psi(n) \to \psi'(n) = V(n)\psi(n) \qquad \bar{\psi}(n) \to \bar{\psi}'(n) = \bar{\psi}(n)V^{\dagger}(n), \tag{3.27}$$

where $V(n)$ is a local $SU(3)_c$ gauge transformation. We are familiar with this problem from the continuum case and the remedy is to define a covariant derivative by using the discrete version of the gauge fields—link variables. We insert them accordingly to form a gauge-invariant action,

$$\begin{aligned} S_F[\psi, \bar{\psi}, U] &= a^4 \sum_{n,m} \bar{\psi}(n) \left[m_q \delta_{n,m} + \frac{1}{2a} \sum_{\mu=1}^{4} \gamma_\mu \left(U_\mu(n)\delta_{n+\hat{\mu},m} - U_\mu^{\dagger}(n-\hat{\mu})\delta_{n-\hat{\mu},m} \right) \right] \psi(m) \\ &= a^4 \sum_{n,m} \bar{\psi}(n) D(n,m) \psi(m), \end{aligned} \tag{3.28}$$

where we have defined the term in the square brackets as the Dirac operator, $D(n,m)$. Using Eq. (3.14), this form can be shown to be gauge invariant.

In order to convince ourselves that this prescription corresponds to the continuum action, consider the Taylor expansions of link variables and fields in the $a \to 0$ limit up to $\mathcal{O}(a)$,

$$U_\mu(n) = 1 + iag_s A_\mu(n) + \mathcal{O}(a^2), \tag{3.29}$$
$$U_\mu^{\dagger}(n-\hat{\mu}) = 1 - iag_s A_\mu(n-\hat{\mu}) + \mathcal{O}(a^2), \tag{3.30}$$
$$A_\mu(n-\hat{\mu}) = A_\mu(n) + \mathcal{O}(a), \tag{3.31}$$
$$\psi(n \pm \hat{\mu}) = \psi(n) + \mathcal{O}(a). \tag{3.32}$$

Inserting the above expressions accordingly, we recover the continuum action,

$$S_F[\psi, \bar{\psi}, U] = a^4 \sum_n \sum_{\mu=1}^{4} \bar{\psi}(n) \left[\gamma_\mu \partial_\mu + i g_s \gamma_\mu A^\mu(n) + m_q \right] \psi(n) + \mathcal{O}(a^2) \tag{3.33}$$

$$\to \int d^4x\, \bar{\psi}(x) \left(\mathcal{D}_\mu + m_q \right) \psi(x) + \mathcal{O}(a^2), \tag{3.34}$$

where $\mathcal{D}_\mu$ is the continuum covariant derivative.

Some properties of the naive fermion action are:

- Discretisation errors start at $\mathcal{O}(a^2)$.
- It is invariant under the global symmetry:

$$\psi(n) \to e^{i\alpha}\psi(n) \quad \bar{\psi}(n) \to \bar{\psi}(n)e^{-i\alpha}, \tag{3.35}$$

where α is just a continous parameter. This symmetry is related to the baryon number conservation and leads to the conservation of the vector current.

• Considering massless fermions, i.e. $m_q = 0$, naive action remains invariant under chiral transformations,

$$\psi(n) \rightarrow e^{i\alpha\gamma_5}\psi(n) \quad \bar{\psi}n \rightarrow \bar{\psi}(n)e^{-i\alpha\gamma_5}, \tag{3.36}$$

as well. It implies that the axial current is conserved, which is in conflict with the continuum Adler-Bell-Jackwin anomaly [11, 12] where the divergence of the axial current is nonzero. Unphysical fermion modes (doublers), turns out to be responsible for the cancellation of the ABJ anomaly [13].

• Naive fermion action has unphysical fermion modes.

3.4.2 Fermion Doubling

Naive discretisation procedure leads to $2^{d=4} = 16$ fermion species in the continuum limit, 15 of which are extra or doublers. Analysing the quark propagator gives an insight to the *fermion doubling* problem. Recall the Dirac operator in Eq. (3.28),

$$D(n, m) = m_q \delta_{n,m} + \frac{1}{2a} \sum_{\mu=1}^{4} \gamma_\mu \left(U_\mu(n)\delta_{n+\hat{\mu},m} - U_\mu^\dagger(n)\delta_{n-\hat{\mu},m} \right). \tag{3.37}$$

Using the exponent definition of the Kronecker delta function,

$$\delta_{n,m} = \frac{1}{|\Lambda|} \sum_{k_\mu} e^{-iak_\mu(n-m)}, \tag{3.38}$$

where $|\Lambda| = N_S^3 N_T$ is the total number of lattice sites, Fourier transform the Dirac operator on a trivial gauge configuration, i.e. $U_\mu = 1$,

$$\tilde{D}(n, m) = \frac{1}{|\Lambda|} \sum_{n,m} \sum_{k_\mu} \left\{ m_q e^{-iak_\mu(n-m)} + \frac{1}{2a} \sum_{\mu=1}^{4} \gamma_\mu \left(e^{-iak_\mu(n+\hat{\mu}-m)} - e^{-iak_\mu(n-\hat{\mu}-m)} \right) \right\}. \tag{3.39}$$

Factoring out an $e^{-iak_\mu(\mathbf{n}-\mathbf{m})}$ and using Eq. (3.38) once again, we get,

$$\tilde{D}(n, m; k) = \delta_{n,m} \left(m_q + \frac{i}{a} \sum_{\mu=1}^{4} \sin(k_\mu a)\gamma_\mu \right) \tag{3.40}$$

$$= \delta_{n,m} D(k) \tag{3.41}$$

where we have also used the Euler's formula for the *sine* function, dropped the unit vector $|\hat{\mu}| = 1$ and defined the term in parentheses as $D(k)$. Multiplying with $D(k)^{-1}$ from right we find the relation,

$$\tilde{D}(k)D(k)^{-1} = \delta_{n,m}, \tag{3.42}$$

where the inverse of the $D(k)$ is,

$$D(k)^{-1} = \frac{m_q - ia^{-1}\sum_\mu \sin(k_\mu a)\gamma_\mu}{m_q^2 + a^{-2}\sum_\mu \sin^2(k_\mu a)}, \tag{3.43}$$

and by inverse Fourier transforming we obtain the quark propagator $D(n, m)^{-1}$,

$$D(n, m)^{-1} = \frac{1}{|\Lambda|}\sum_{k_\mu} D^{-1}(k)e^{-iak_\mu(n-m)}. \tag{3.44}$$

Fermion doubling shows itself in the momentum space propagator in Eq. (3.43). Written down for massless fermions,

$$D(k)^{-1} = \frac{-ia^{-1}\sum_\mu \sin(k_\mu a)\gamma_\mu}{a^{-2}\sum_\mu \sin^2(k_\mu a)}, \tag{3.45}$$

we see that the term in the denominator has a pole for $k_\mu = (0, 0, 0, 0)$, corresponding to physical fermions, along with 15 more due to the periodicity of the sine function. Extra poles, or unphysical fermions, lie at $k_\mu = \pi$ and 0, i.e. $k_\mu = (\pi/a, 0, 0, 0)$, $(0, \pi/a, 0, 0), \ldots, (\pi/a, \pi/a, \pi/a, \pi/a)$.

3.4.3 Wilson Fermions

It is necessary to remove the extra modes to have a reliable lattice theory. Solution, as suggested by Wilson [14], is to add an extra operator to the naive action which does not change its continuum limit. The relevant term is a dimension-5 Laplace operator,

$$ar\hat{\Box} = \frac{ar}{2a^2}\sum_{\mu=1}^{4}\left(U_\mu(n)\delta_{n+\hat{\mu},m} - 2\delta_{n,m} + U_\mu^\dagger(n-\hat{\mu})\delta_{n-\hat{\mu},m}\right), \tag{3.46}$$

where the constant a ensures that the *Wilson term* vanishes as $a \to 0$. With the addition of this term, Dirac operator becomes,

$$D(n,m) = \left(m_q + \frac{4r}{a}\right)\delta_{n,m} - \frac{1}{2a}\sum_{\mu=1}^{4}\left[\left(r-\gamma_\mu\right)U_\mu(n)\delta_{n+\hat{\mu},m} + \left(r+\gamma_\mu\right)U_\mu^\dagger(n)\delta_{n-\hat{\mu},m}\right]$$
$$= \delta_{n,m} - \kappa\sum_{\mu=1}^{4}\left[\left(r-\gamma_\mu\right)U_\mu(n)\delta_{n+\hat{\mu},m} + \left(r+\gamma_\mu\right)U_\mu^\dagger(n)\delta_{n-\hat{\mu},m}\right], \tag{3.47}$$

where we have introduced the κ parameter,

$$\kappa = \frac{1}{2(m_q a + 4r)}, \tag{3.48}$$

and by rescaling the fermion fields as,

$$\psi \equiv \sqrt{m_q a + 4r}\,\psi = \psi/\sqrt{2\kappa}, \tag{3.49}$$

doubler-free Wilson fermion action is written as,

$$S_F^W[\psi,\bar{\psi},U] = a^4\sum_{\mathbf{n},\mathbf{m}}\bar{\psi}(n)D(n,m)\psi(m),$$
$$D(n,m) = \delta_{n,m} - \kappa\sum_{\mu=1}^{4}\left[\left(r-\gamma_\mu\right)U_\mu(n)\delta_{n+\hat{\mu},m} + \left(r+\gamma_\mu\right)U_\mu^\dagger(n)\delta_{n-\hat{\mu},m}\right]. \tag{3.50}$$

We may follow a similar procedure to that of the previous section and analyse the momentum-space propagator. Wilson term transforms into a cosine term,

$$\tilde{D}_W(k)^{-1} = a^{-1}\left[1 - 2\kappa\left(i\sum_{\mu=1}^{4}\gamma_\mu \sin(k_\mu a) + \sum_{\mu=1}^{4} r\cos(k_\mu a)\right)\right], \tag{3.51}$$

lifting the doublers such that, they acquire masses of $\mathcal{O}(1/a)$ for each $k_\mu a = \pi$ component, so, they get heavy and decouple in the continuum limit as $a \to 0$.

Some properties of the Wilson fermions are:

• Doublers are removed by adding a higher dimensional term to the naive action. They acquire a mass of $\mathcal{O}(1/a)$ and decouple in the continuum limit. ABJ anomaly is restored [15].
• We have picked up a $4r/a$ term along the way which acts as a mass term (see Eq. (3.47)) and breaks the chiral symmetry explicitly even in the $m_q \to 0$ limit. As a consequence, for example, the axial Ward-Takahashi identity receives $\mathcal{O}(a)$ corrections. Such quantities are used to improve the lattice actions and operators by matching the lattice results with the continuum [16].

There is, unfortunately, no remedy for the explicit chiral symmetry breaking in Wilson's formulation. One has to consider other formulations, e.g. staggered [17],

overlap [18] or domain-wall [19] fermions provide actions that preserve chiral symmetry and are doubler-free while, however, introducing other setbacks.
• Global

$$\psi(n) \rightarrow e^{i\alpha}\psi(n), \quad \bar{\psi}n \rightarrow \bar{\psi}ne^{-i\alpha}, \tag{3.52}$$

symmetry is still preserved. One can derive the corresponding conserved non-local vector current via Noether procedure as,

$$\mathcal{V}_\mu(x) = \frac{1}{2}[\bar{q}(x+\mu)U_\mu^\dagger(r+\gamma_\mu)q(x) - \bar{q}(x)U_\mu(r-\gamma_\mu)q(x+\mu)]. \tag{3.53}$$

This current can be employed without renormalisation since it is conserved.
• There is a one-to-one correspondence between the continuum and lattice operators one can form. We simply formulate the observables in continuum language in Sect. 4.1 and estimate their values by ensemble averages.
• $\mathcal{O}(a^2)$ discretisation errors of the naive action is demoted to $\mathcal{O}(a)$ due to the additional Wilson term, which calls for an improvement.

3.4.4 Improved Action

We summarise how an $\mathcal{O}(a)$-improvement is achieved via Symanzik's improvement program [20–25]. We have mentioned a few times before that we can add higher dimension terms to the lattice action which serve to remove the discretisation artefacts as long as the continuum form of the action is preserved. Formally, this statement may be understood from an effective theory approach [22, 23] with the action,

$$S_{Cont}(x) = \int d^4x \left\{\mathcal{L}_{Lat}(x) + a\mathcal{L}_1(x) + a^2\mathcal{L}_2(x) + \ldots\right\}, \tag{3.54}$$

where $\mathcal{L}_{Lat}(x)$ is the dimension-4 lattice Lagrangian density and $\mathcal{L}_i(x)$ include $4+i$ dimensional local operators. For $\mathcal{O}(a)$-improvement one should add counter terms that cancel $\mathcal{L}_1(x)$. Respecting the underlying symmetries, there appears to be five dimension-5 operators,

$$\begin{aligned}
\hat{\mathcal{O}}_1 &= \bar{\psi}\sigma_{\mu\nu}F_{\mu\nu}\psi, \\
\hat{\mathcal{O}}_2 &= \bar{\psi}D_\mu D_\mu\psi + \bar{\psi}\overleftarrow{D}_\mu\overleftarrow{D}_\mu\psi, \\
\hat{\mathcal{O}}_3 &= m_q \mathrm{Tr}\left[F_{\mu\nu}F_{\mu\nu}\right], \\
\hat{\mathcal{O}}_4 &= m_q\left(\bar{\psi}\gamma_\mu D_\mu\psi + \bar{\psi}\overleftarrow{D}_\mu\gamma_\mu\psi\right), \\
\hat{\mathcal{O}}_5 &= m_q^2\bar{\psi}\psi,
\end{aligned}$$

that would construct $\mathcal{L}_1(x)$, where $\hat{\mathcal{O}}_2$ and $\hat{\mathcal{O}}_4$ gets eliminated by use of field equations and $\hat{\mathcal{O}}_3$ and $\hat{\mathcal{O}}_5$ are absorbed into bare coupling and mass renormalisation factors [16]. So the counter term depends only on $\hat{\mathcal{O}}_1$ and the improved action is written down as [26],

$$S_F^C = S_F^W + a\kappa_q a^4 \sum_n \sum_{\mu<\nu} c_{SW} \bar{\psi}(n) \frac{1}{2} \sigma_{\mu\nu} \hat{F}_{\mu\nu}(n) \psi(n), \tag{3.55}$$

where S_F^W is the Wilson action in Eq. (3.50), c_{SW} is the Sheikholeslami-Wohlert term whose value should be tuned to eliminate the $\mathcal{O}(a)$ term and we have separated the a^4 term that is associated with the sum over n which is the discrete version of the four-dimensional integral. Here, $\sigma_{\mu\nu} = \frac{1}{2}[\gamma_\mu, \gamma_\nu]$, κ_q is defined in Eq. (3.48) and $\hat{F}_{\mu\nu}(n)$ is the gluon field strength tensor shown in Fig. 3.3 and defined as,

$$F_{\mu\nu}(n) = \frac{1}{8i} \left(Q_{\mu\nu}(n) - Q_{\mu\nu}^\dagger(n) \right) \tag{3.56}$$

$$Q_{\mu\nu}(n) = U_{\mu,\nu}(n) + U_{\nu,-\mu}(n) + U_{-\mu,-\nu}(n) + U_{-\nu,\mu}(n), \tag{3.57}$$

where $U_{\mu,\nu}$ are 1×1 plaquettes.

Improvement coefficient is $c_{SW} = 1$, in the tree-level approximation and determined in 1-loop order [27, 28] as well. In order to fully remove the $\mathcal{O}(a)$ artefacts, however, one needs a non-perturbative improvement which is usually done by tuning c_{SW} to reproduce the axial Ward-Takahashi identity up to $\mathcal{O}(a^2)$ [16]. PACS-CS Collaboration follows the non-perturbative method and determines the improvement coefficient as $c_{SW} = 1.715$ [29].

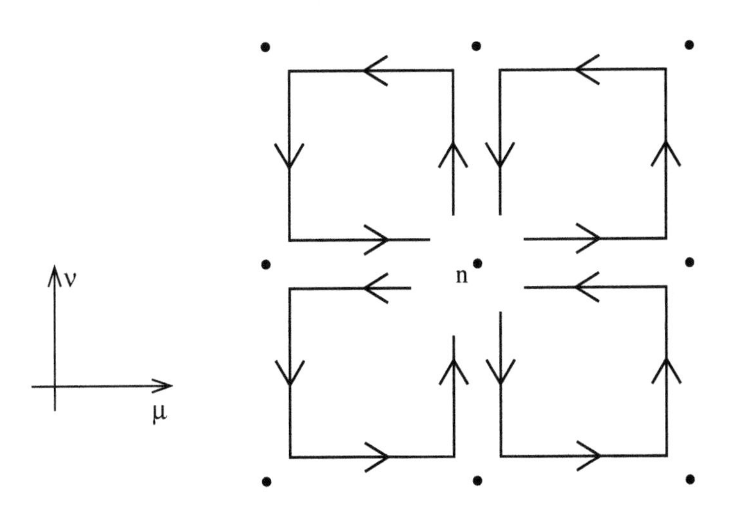

Fig. 3.3 The clover term. Sum of plaquettes in the μ-ν plane. Compare with Eq. (3.15) and Fig. 3.1

3.5 Workflow

We sketch some aspects of a typical application of lattice QCD to the computation of a generic zero-temperature, zero-density hadronic observable. Details are discussed in length in Chap. 4. As a first glimpse on how to calculate physical quantities via lattice simulations, let us consider a generic correlation function of the form,

$$\langle\hat{\mathcal{O}}\rangle = \frac{1}{Z}\int \mathcal{D}[U]\, e^{-S_G[U]}\mathcal{D}\left[\bar{\psi}, \psi\right] e^{-S_F[\bar{\psi},\psi,U]}\mathcal{O}\left[\bar{\psi}, \psi\right], \tag{3.58}$$

written in Feynman path integral formalism in Euclidean space-time. LHS is in operator language such that $\hat{\mathcal{O}}$ is a combination of operators relevant to the desired quantity and RHS is in functional form with the integral measures over gauge and fermion fields defined by,

$$\mathcal{D}[U] = \Pi_n \Pi_{\mu=1}^4 \mathrm{d}U_\mu(n), \quad \mathcal{D}\left[\bar{\psi}, \psi\right] = \Pi_n \mathrm{d}\bar{\psi}(n)\mathrm{d}\psi(n). \tag{3.59}$$

S_G and S_F are Euclidean gauge and fermion actions as defined in Sects. 3.3 and 3.4 and the partition function is defined as,

$$Z = \int \mathcal{D}[U]\,\mathcal{D}\left[\bar{\psi}, \psi\right] e^{-(S_G+S_F)}. \tag{3.60}$$

Note that switching from a Minkowski description to Euclidean one by a Wick rotation, i.e. $t \to -i\tau$ is a crucial point and has two advantages: (i) We exploit a connection to statistical mechanics and use methods such as Monte Carlo integration and (ii) the exponents are now exponential decays, e.g. e^{-S_G}, in contrast to the oscillating function of Minkowski formulation, e^{-iS_G}, which stabilises the Monte Carlo simulations.

Fermion fields are represented by Grassmann numbers since they obey Fermi-Dirac statistics. Following the integration rules of Grassmann numbers, integration of the fermion action reduces to a fermion determinant form [30, 31],

$$I_F = \int \mathcal{D}\left[\bar{\psi}, \psi\right] e^{\int d^4x \bar{\psi} D \psi} = \det[D], \tag{3.61}$$

where D is the corresponding Dirac operator of the chosen action. Now, our description takes the form,

$$\langle\hat{\mathcal{O}}\rangle = \frac{1}{Z}\int \mathcal{D}[U]\, e^{-S_G[U]}\det[D]\,\mathcal{O}\left[\bar{\psi}, \psi\right], \tag{3.62}$$

where there will be a fermion determinant for each quark flavor that one considers in application, e.g. $\det[D_u]\det[D_d]\det[D_s]$ for three flavors. Fermion determinant plays a crucial role in including the sea-quark effects in simulations since, although not immediately clear here, it represents the fermion loops in the vacuum.

Considering hadronic observables, for instance, operators of the correlation function depend on combinations of quark bilinears composed of quarks fields and gamma matrices, Γ. Wick contraction translates quark fields into quark propagators, $S_q(n, m) = D^{-1}$, which is calculated numerically by inverting the Dirac matrix for each color and Dirac component of the quark. Hadronic correlation function then assumes the form,

$$\langle \hat{O} \rangle = \frac{1}{Z} \int \mathcal{D}[U] e^{-S_G[U]} \det[D] \, \mathrm{Tr}\left[\Gamma S_{q1}(n,m) \Gamma S_{q2}(n,m) \ldots\right]. \tag{3.63}$$

In order to evaluate the integral over the gauge fields we exploit the resemblance of this form to the one in statistical mechanics systems: Consider a spin system whose energy is given by a functional, $E(s)$, of its spins. Broadly speaking, expectation value of some observable, O, is then estimated by summing over its value obtained on all possible configurations of the system,

$$\langle O \rangle = \frac{1}{Z} \sum_{s} e^{-\beta E(s)} O(s), \tag{3.64}$$

where Z is the partition function of the system, $\beta = 1/k_B T$ is defined in terms of Boltzmann constant, k_B and temperature, T, and $e^{-\beta E(s)}$ is the Boltzmann factor. In our case, $e^{-S_G[U]} e^{\log(\det[D])}$ term assumes the role of the Boltzmann factor as a weight function and the value of the observable is simply given by a sum over all possible configurations of the QCD vacuum, so-called gauge ensembles. In practice, however, there may be infinitely many QCD configurations so the observables are estimated by averaging over a sufficient number of ensembles,

$$\langle \hat{O} \rangle = \frac{\int \mathcal{D}[U] e^{-S_G[U]} \det[D] \, \mathcal{O}\left[\bar{\psi}, \psi\right]}{\int \mathcal{D}[U] e^{-S_G[U]} \det[D]} = \lim_{N \to \infty} \frac{1}{N} \sum_{n=1}^{N} \mathcal{O}\left[\bar{\psi}, \psi\right]. \tag{3.65}$$

Although the underlying idea is simple, generating reliable gauge configurations is a formidable task usually tackled by collaborations with access to vast computational resources. Gauge configurations are generated by Monte Carlo simulations with several algorithms that are being employed. We do not cover the details but refer the reader to other pedagogical sources such as the lecture notes by Gattringer and Lang [32]. We do note, however, that gauge configuration generation is a highly technical field, also from a computer science point of view, which challenges high-performance computing centers, even with dedicated supercomputers to the task in some cases [33].

Historically, fermion determinant, hence the sea-quark effects, was ignored by setting $\det[D] = 1$, which is known as the quenched approximation. Proceeding by this method is still used as a first means of exploratory calculations, however, one needs full QCD calculations, which include the determinant, to draw realistic conclusions from a lattice approach.

We may form a simple recipe to summarise the necessary steps to evaluate hadronic observables on the lattice as:

- *Generate gauge configurations*: Note that we do not generate any for this work but employ the publicly available ones. Details are given in Sect. 4.2.1.
- *Construct the correlation functions via operators corresponding to the hadronic observable in question and re-express it in terms of quark propagators*: We cover the details in Sects. 4.1.2.1 and 4.1.2.2 and some technical details about propagator calculation are given in Sects. 4.2.3 and 4.2.5
- *Evaluate the ensemble average and estimate the observable*: We discuss an array of results in Chap. 5.

References

1. R.P. Feynman, Space-time approach to non-relativistic quantum mechanics. Rev. Mod. Phys. **20**: 367–387 (1948). https://doi.org/10.1103/RevModPhys.20.367
2. K. Symanzik, Schrödinger representation and casimir effect in renormalizable quantum field theory. Nuclear Physics B **190**(1): 1–44 (1981). ISSN 0550-3213. https://doi.org/10.1016/0550-3213(81)90482-X. URL http://www.sciencedirect.com/science/article/pii/055032138190482X. Volume B190 [FS3] No. 2 To Follow in Approximately Two Months
3. M. Lüscher, Schrödinger representation in quantum field theory. Nuclear Phys. B **254**(0): 52–57 (1985). ISSN 0550-3213. https://doi.org/10.1016/0550-3213(85)90210-X. http://www.sciencedirect.com/science/article/pii/055032138590210X
4. M. Luscher, R. Narayanan, P. Weisz, U. Wolff, The Schrodinger functional: a renormalizable probe for non-Abelian gauge theories. Nucl. Phys. **B384**:168–228 (1992). https://doi.org/10.1016/0550-3213(92)90466-O
5. S. Sint, On the Schrodinger functional in QCD. Nucl. Phys. **B421**: 135–158 (1994). https://doi.org/10.1016/0550-3213(94)90228-3
6. S. Sint, One loop renormalization of the QCD Schrodinger functional. Nucl. Phys. **B451**: 416–444 (1995). https://doi.org/10.1016/0550-3213(95)00352-S
7. S. Aoki, K.-I. Ishikawa, N. Ishizuka, T. Izubuchi, D. Kadoh, K. Kanaya, Y. Kuramashi, Y. Namekawa, M. Okawa, Y. Taniguchi, A. Ukawa, N. Ukita, T. Yoshie, 2+1 flavor lattice QCD toward the physical point. Phys. Rev. **D79**: 034503 (2009). https://doi.org/10.1103/PhysRevD.79.034503
8. S. Aoki et al., Comparative study of full QCD hadron spectrum and static quark potential with improved actions. Phys. Rev. **D60**: 114508 (1999). https://doi.org/10.1103/PhysRevD.60.114508
9. Y. Iwasaki, Renormalization group analysis of lattice theories and improved lattice action. II–four-dimensional non-abelian SU(N) gauge model (2011). arXiv:1111.7054
10. M. Luscher, P. Weisz, On-shell improved lattice gauge theories. Commun. Math. Phys. **97**:59 (1985). https://doi.org/10.1007/BF01206178. [Erratum: Commun. Math. Phys. 98, 433 (1985)]
11. J.S. Bell, R. Jackiw, A PCAC puzzle: pi0 –> gamma gamma in the sigma model. Nuovo Cim. **A60**: 47–61 (1969). https://doi.org/10.1007/BF02823296

12. S.L. Adler, Axial-vector vertex in spinor electrodynamics. Phys. Rev. **177**: 2426–2438 (1969). https://doi.org/10.1103/PhysRev.177.2426
13. R. Gupta. Introduction to lattice QCD: Course, in *Probing the standard model of particle interactions. Proceedings, summer school in theoretical physics*, NATO Advanced Study Institute, 68th session, Les Houches, France, July 28–September 5, 1997. Pt. 1, 2, pp. 83–219 (1997). http://alice.cern.ch/format/showfull?sysnb=0284452
14. K.G. Wilson, New phenomena in subnuclear physics: Part a, in *New phenomena in subnuclear physics: part A*, ed. by A. Zichichi (Springer, US, Boston, MA, 1977). ISBN 978-1-4613-4208-3. https://doi.org/10.1007/978-1-4613-4208-3_6
15. L.H. Karsten, J. Smith, Lattice fermions: species doubling, chiral invariance and the triangle anomaly. Nuclear Phys. B **183**(1): 103–140 (1981). ISSN 0550-3213. https://doi.org/10.1016/0550-3213(81)90549-6. http://www.sciencedirect.com/science/article/pii/0550321381905496
16. M. Lüscher, S. Sint, R. Sommer, P. Weisz, Chiral symmetry and o(a) improvement in lattice QCD. Nuclear Phys. B **478**(1): 365–397 (1996). ISSN 0550-3213. https://doi.org/10.1016/0550-3213(96)00378-1. http://www.sciencedirect.com/science/article/pii/0550321396003781
17. J.B. Kogut, L. Susskind, Hamiltonian formulation of Wilson's lattice gauge theories. Phys. Rev. **D11**: 395 (1975). https://doi.org/10.1103/PhysRevD.11.395
18. Herbert Neuberger, Exactly massless quarks on the lattice. Phys. Lett. B **417**, 141–144 (1998)
19. D.B. Kaplan, A method for simulating chiral fermions on the lattice. Phys. Lett. **B288**: 342–347 (1992). https://doi.org/10.1016/0370-2693(92)91112-M
20. G. Curci, P. Menotti, G. Paffuti, Symanzik's improved lagrangian for lattice gauge theory. Phys. Lett. B **130**(3): 205–208 (1983). ISSN 0370-2693. https://doi.org/10.1016/0370-2693(83)91043-2. http://www.sciencedirect.com/science/article/pii/0370269383910432
21. K. Symanzik. Some topics in quantum field theory, in *Mathematical Problems in Theoretical Physics: Proceedings of the VIth International Conference on Mathematical Physics Berlin (West) ed. by R. Schrader, R. Seiler, D.A. Uhlenbrock, August 11–20, 1981* (Springer, Heidelberg, 1982), pp. 47–58. ISBN 978-3-540-38982-8. https://doi.org/10.1007/3-540-11192-1_11
22. K. Symanzik, Continuum limit and improved action in lattice theories. Nuclear Phys. B **226**(1): 187–204 (1983a). ISSN 0550-3213. https://doi.org/10.1016/0550-3213(83)90468-6. http://www.sciencedirect.com/science/article/pii/0550321383904686
23. K. Symanzik, Continuum limit and improved action in lattice theories. Nuclear Phys. B **226**(1): 205–227 (1983b). ISSN 0550-3213. https://doi.org/10.1016/0550-3213(83)90469-8. http://www.sciencedirect.com/science/article/pii/0550321383904698
24. P. Weisz, Continuum limit improved lattice action for pure yang-mills theory (i). Nuclear Phys. B **212** (1): 1–17 (1983). ISSN 0550-3213. https://doi.org/10.1016/0550-3213(83)90595-3. http://www.sciencedirect.com/science/article/pii/0550321383905953
25. P. Weisz, R. Wohlert, Continuum limit improved lattice action for pure yang-mills theory (ii). Nuclear Phys. B **236**(2):397–422 (1984). ISSN 0550-3213. https://doi.org/10.1016/0550-3213(84)90543-1. http://www.sciencedirect.com/science/article/pii/0550321384905431
26. B. Sheikholeslami, R. Wohlert, Improved continuum limit lattice action for QCD with Wilson Fermions. Nucl. Phys. **B259**: 572 (1985). https://doi.org/10.1016/0550-3213(85)90002-1
27. M. Lüscher, P. Weisz, O(a) improvement of the axial current in lattice QCD to one-loop order of perturbation theory. Nuclear Phys. B **479**(1): 429–458 (1996). ISSN 0550-3213. https://doi.org/10.1016/0550-3213(96)00448-8. http://www.sciencedirect.com/science/article/pii/0550321396004488
28. R. Wohlert, Improved continuum limit lattice action for quarks. *DESY-87-069* (1987)
29. S. Aoki et al. Nonperturbative O(a) improvement of the Wilson quark action with the RG-improved gauge action using the Schrodinger functional method. Phys. Rev. **D73**: 034501 (2006). https://doi.org/10.1103/PhysRevD.73.034501
30. P.T. Matthews, A. Salam, The green's functions of quantised fields. Il Nuovo Cimento (1943–1954) **12**(4): 563–565 (1954). ISSN 1827-6121. https://doi.org/10.1007/BF02781302

31. P.T. Matthews, A. Salam, Propagators of quantized field. Il Nuovo Cimento (1955–1965) **2**(1): 120–134 (1955). ISSN 1827-6121. https://doi.org/10.1007/BF02856011
32. C. Gattringer, C.B. Lang, *Quantum chromodynamics on the lattice*, *Lecture Notes in Physics*, vol. 788, 1 edn. (Springer, Heidelberg, 2010). https://doi.org/10.1007/978-3-642-01850-3
33. D. Chen, N.H. Christ, C. Cristian, Z. Dong, A. Gara, K. Garg, B. Joo, C. Kim, L. Levkova, X. Liao, R.D. Mawhinney, S. Ohta, T. Wettig, Qcdoc: A 10-teraflops scale computer for lattice QCD. Nuclear Phys. B - Proc. Suppl. **94**(1): 825–832 (2001). ISSN 0920-5632. https://doi.org/10.1016/S0920-5632(01)01014-3. http://www.sciencedirect.com/science/article/pii/S0920563201010143

Part II
Formalism and Results

Chapter 4
Theoretical Formalism and Simulation Setup

Abstract This chapter consists of two parts where we, first, discuss the essential theoretical formalism to investigate the observables we are interested in. Mainly, we provide the formulations to extract the mass and the electromagnetic form factors of spin-1/2 and spin-3/2 baryons. A brief account on data analysis is also given. Second part focuses on the technical, computaional aspects of the lattice method where we detail our setup. Information about the gauge configurations, parameter tunings, propagator inversions and statistical improvements are all given in this part.

Keywords Hadron mass · Electromagnetic form factor
Spin-1/2 and spin-3/2 baryons · Lattice simulations · Parameter tuning

4.1 Theoretical Formalism

4.1.1 Hadron Masses

Mass of a hadron is one of the crucial properties that we use to identify a particular hadron along with its quantum numbers. Hadron spectrum calculations have been (and still are) central to our understanding of the strong interactions. In the continuum quantum field theory language, the mass of a particle is encoded into its two-point correlation function which describes the particle freely propagating along the time direction. It is better to write the two-point correlation function as a spectral decomposition to make the discussion more clear, so let us write down the correspondence for a positive parity spin-1/2 baryon as an example and continue with the lattice method to extract the mass.

4.1.1.1 Spectral Decomposition

We define the two-point correlation function of a spin-1/2 baryon projected to a definite momentum as,

K. U. Can, *Electromagnetic Form Factors of Charmed Baryons in Lattice QCD*,
Springer Theses, https://doi.org/10.1007/978-981-10-8995-4_4

$$\langle G^{\mathcal{BB}}(t;\mathbf{p};\Gamma)\rangle = \sum_{x} e^{-i\mathbf{p}\cdot\mathbf{x}}\Gamma^{\beta\alpha}\langle 0|\mathcal{T}\{\chi^{\alpha}_{\mathcal{B}}(\mathbf{x},t)\bar{\chi}^{\beta}_{\mathcal{B}}(0)\}|0\rangle, \quad (4.1)$$

where $\chi_{\mathcal{B}}$ is the interpolating field of the baryon and α and β are Dirac indices. Inserting the identity operator as a complete set of states, $\sum_{\mathcal{B},s}|\mathcal{B}(p,s)\rangle\langle\mathcal{B}(p,s)|$ we end up with a sum over all possible states that correspond to the quantum numbers defined by the $\chi_{\mathcal{B}}$ interpolating field,

$$\langle G^{\mathcal{BB}}(t;\mathbf{p};\Gamma)\rangle = \sum_{\mathcal{B},s} e^{-E_{\mathcal{B}}(\mathbf{p})t}\Gamma^{\beta\alpha}\langle 0|\chi^{\alpha}_{\mathcal{B}}(\mathbf{p})|\mathcal{B}(p,s)\rangle\langle\mathcal{B}(p,s)|\bar{\chi}^{\beta}_{\mathcal{B}}(0)|0\rangle \quad (4.2)$$

$$= A_0(\mathbf{p})e^{-E_0(\mathbf{p})t} + A_1(\mathbf{p})e^{-E_1(\mathbf{p})\mathbf{t}} + \ldots, \quad (4.3)$$

where we have factored out the time dependence of the fields, acted it upon the states properly and have collected them into the functions $A_n(\mathbf{p})$ for simplicity. E_n corresponds to the energy of the respective state starting from the $n = 0$ ground state. Since we are interested in the mass of the lowest-lying state we take the large Euclidean time limit, $t \gg a$, so that the ground state $\mathcal{B}$ dominates and the correlation function reduces to the first term of Eq. (4.3),

$$\langle G^{\mathcal{BB}}(t;\mathbf{p};\Gamma)\rangle = \sum_{s} e^{-E_{\mathcal{B}}(\mathbf{p})t}\Gamma^{\beta\alpha}\langle 0|\chi^{\alpha}_{\mathcal{B}}(p)|\mathcal{B}(p,s)\rangle\langle\mathcal{B}(p,s)|\bar{\chi}^{\beta}_{\mathcal{B}}(0)|0\rangle, \quad (4.4)$$

where we have now written the spin sum explicitly and identified the ground state with definite momentum. The overlap between the interpolating field and the physical state is defined as,

$$\langle 0|\chi^{\alpha}_{\mathcal{B}}(0)|\mathcal{B}(\mathbf{p},s)\rangle = Z_{\mathcal{B}}\sqrt{\frac{M_{\mathcal{B}}}{E_{\mathcal{B}}(p)}}u(p,s), \quad (4.5)$$

where $Z_{\mathcal{B}}$ is the overlap factor, $M_{\mathcal{B}}$ is the mass of the baryon and $u(p,s)$ is the Dirac spinor. Using the Dirac spinor sum,

$$\sum_{s} u(p,s)\bar{u}(p,s) = \frac{\gamma_{\mu}p^{\mu} + M_{\mathcal{B}}}{2M_{\mathcal{B}}}, \quad (4.6)$$

and inserting Eq. (4.5) into Eq. (4.4); the two-point function takes the form:

$$\langle G^{\mathcal{BB}}(t;\mathbf{p};\Gamma)\rangle = |Z_{\mathcal{B}}|^2\frac{M_{\mathcal{B}}}{E_{\mathcal{B}}(\mathbf{p})}e^{-E_{\mathcal{B}}(\mathbf{p})t}\mathrm{Tr}[\Gamma\frac{\gamma_{\mu}p^{\mu} + M_{\mathcal{B}}}{2M_{\mathcal{B}}}]. \quad (4.7)$$

Finally using the definitions of the Γ matrices as,

$$\Gamma_i = \frac{1}{2}\begin{pmatrix} \sigma_i & 0 \\ 0 & 0 \end{pmatrix}, \qquad \Gamma_4 = \frac{1}{2}\begin{pmatrix} I & 0 \\ 0 & 0 \end{pmatrix}, \tag{4.8}$$

we find that only the Γ_4 component survives and setting the three momentum to $\mathbf{p} = (0, 0, 0)$ reduces the expression to,

$$\langle G^{\mathcal{BB}}(t; \mathbf{0}; \Gamma_4)\rangle = |Z_{\mathcal{B}}|^2 \, e^{-M_{\mathcal{B}}t}. \tag{4.9}$$

This quick derivation immediately gives us the functional form that we will use in regression analysis (described in Sect. 4.1.3) of the two-point correlation function and extract the mass of $\mathcal{B}$.

4.1.1.2 Lattice Method

Once we compute the lattice two-point correlation function it is a simple procedure to extract the mass of the baryon. However there is an important issue to be aware of, which is evident if we take a closer look to the Eq. (4.2),

$$\langle G^{\mathcal{BB}}(t; \mathbf{p}; \Gamma)\rangle = A_0(\mathbf{p})e^{-E_0(\mathbf{p})\mathbf{t}} + A_1(\mathbf{p})e^{-E_1(\mathbf{p})t} + \dots. \tag{4.10}$$

The sum over all possible states indicates that not only the ground state of the baryon but also all the possible excited states or many-particle states that correspond to the same quantum numbers are included in the correlation function. Exponential terms, on the other hand, ensure that all the states decay with respect to the elapsed time. Rearranging Eq. (4.10) for the ground state with zero-momentum we have,

$$\langle G^{\mathcal{BB}}(t; \mathbf{0}; \Gamma)\rangle = Ae^{-M_{\mathcal{B}}t}(1 + \mathcal{O}(e^{-\Delta E t}) + \dots), \tag{4.11}$$

where ΔE is the energy gap between the ground state and the first excited state. Notice that the contributions of the excited states fall off faster compared to the ground state and ideally the larger time slices should contain the information from the ground state only. This simple observation leads us to form a quantity called *effective mass*,

$$m_{\text{eff}}(t + \frac{1}{2}) = \frac{G^{\mathcal{BB}}(t)}{G^{\mathcal{BB}}(t+1)}, \tag{4.12}$$

which basically allows us to pinpoint the time slice that the ground state signal saturates and guides our analysis of the lattice two-point correlation function. The analysis has a few steps to follow:

1. Calculate the effective mass and plot it with respect to time, t,
2. Pinpoint the time slice, t_i, where the curve starts to form a plateau,
3. Define the fit window $[t_i, t_f]$,
4. Fit $G^{\mathcal{BB}}(t; \mathbf{0}; \Gamma)$ to Eq. (4.9) and extract the mass.

The crucial point is to identify the plateau region which is the key to a reliable mass extraction. Usually one chooses the initial time slice by intuition or, more technically, by shifting the initial choice by a few steps and picking the one having the best response to a goodness of fit analysis. It is good practice to define the fit region until the point where the signal is deemed to be lost [1]. The upper limit of the fit window in our case is limited to the half of the temporal lattice extent $N_t/2$ since the backward propagating (opposite parity) states interfere for the $t > N_t/2$ time slices on periodic boundary lattices. Adding extra terms to account for backward propagating states is a reasonable option however further complicating the fit function would decrease the quality of the fit and is not necessary for our analysis.

4.1.2 Form Factors

Electromagnetic form factors of baryons can be calculated through their matrix elements of the electromagnetic vector current $V_\mu = \sum_q e_q \bar{q}(x)\gamma_\mu q(x)$, where q runs over the quark content of the baryon in consideration. In the following sections we write down the transition matrix elements for spin-1/2 $\rightarrow$ spin-1/2 and spin-3/2 $\rightarrow$ spin-3/2 transitions and show how to identify and extract the form factors from lattice QCD simulations.

4.1.2.1 Spin-1/2 Elastic Form Factors

We have written down the electromagnetic transition matrix element for the spin$-1/2 \rightarrow$ spin$-1/2$ transition in Sect. 2.4 when we discussed the hadron structure but let us recall it again for completeness. Matrix element is written in the following form

$$\langle \mathcal{B}(p', s')|V_\mu|\mathcal{B}(p, s)\rangle = \bar{u}(p', s')\left[\gamma_\mu F_{1,\mathcal{B}}(q^2) + i\frac{\sigma_{\mu\nu}q^\nu}{2M_\mathcal{B}}F_{2,\mathcal{B}}(q^2)\right]u(p, s), \tag{4.13}$$

where $q_\mu = p'_\mu - p_\mu$ is the transferred four-momentum and $\sigma_{\mu\nu} = \frac{1}{2}\{\gamma_\mu, \gamma_\nu\}$. Here $u(p)$ denotes the Dirac spinor for the baryon with four-momentum p^μ and mass $M_\mathcal{B}$. The Dirac, $F_{1,\mathcal{B}}(q^2)$, and Pauli, $F_{2,\mathcal{B}}(q^2)$, form factors are related to the Sachs electric and magnetic form factors by the relations,

$$G_{E,\mathcal{B}}(q^2) = F_{1,\mathcal{B}}(q^2) + \frac{q^2}{4M_{\mathcal{B}}^2} F_{2,\mathcal{B}}(q^2), \quad (4.14)$$

$$G_{M,\mathcal{B}}(q^2) = F_{1,\mathcal{B}}(q^2) + F_{2,\mathcal{B}}(q^2). \quad (4.15)$$

We follow the method outlined in Ref. [2] which is employed to extract the nucleon electromagnetic form factors via computing the matrix element in Eq. 4.13 using the following ratio,

$$R(t_2, t_1; \mathbf{p}', \mathbf{p}; \Gamma; \mu) = \frac{\langle G^{\mathcal{B}\mathcal{V}_\mu\mathcal{B}}(t_2, t_1; \mathbf{p}', \mathbf{p}; \Gamma)\rangle}{\langle G^{\mathcal{B}\mathcal{B}}(t_2; \mathbf{p}'; \Gamma_4)\rangle} \times \left[\frac{\langle G^{\mathcal{B}\mathcal{B}}(t_2 - t_1; \mathbf{p}; \Gamma_4)\rangle\langle G^{\mathcal{B}\mathcal{B}}(t_1; \mathbf{p}'; \Gamma_4)\rangle\langle G^{\mathcal{B}\mathcal{B}}(t_2; \mathbf{p}'; \Gamma_4)\rangle}{\langle G^{\mathcal{B}\mathcal{B}}(t_2 - t_1; \mathbf{p}'; \Gamma_4)\rangle\langle G^{\mathcal{B}\mathcal{B}}(t_1; \mathbf{p}; \Gamma_4)\rangle\langle G^{\mathcal{B}\mathcal{B}}(t_2; \mathbf{p}; \Gamma_4)\rangle}\right]^{1/2}, \quad (4.16)$$

where the baryonic two-point and three-point correlation functions are respectively defined as,

$$\langle G^{\mathcal{B}\mathcal{B}}(t; \mathbf{p}; \Gamma_4)\rangle \equiv \sum_{\mathbf{x}} e^{-i\mathbf{p}\cdot\mathbf{x}} \Gamma_4^{\alpha\alpha'} \langle \text{vac}|\mathcal{T}[\chi_{\mathcal{B}}^{\alpha}(x)\bar{\chi}_{\mathcal{B}}^{\alpha'}(0)]|\text{vac}\rangle, \quad (4.17)$$

$$\langle G^{\mathcal{B}\mathcal{V}_\mu\mathcal{B}}(t_2, t_1; \mathbf{p}', \mathbf{p}; \Gamma)\rangle \equiv -i \sum_{\mathbf{x}_2, \mathbf{x}_1} e^{-i\mathbf{p}\cdot\mathbf{x}_2} e^{i\mathbf{q}\cdot\mathbf{x}_1} \Gamma^{\alpha\alpha'} \times \langle \text{vac}|\mathcal{T}[\chi_{\mathcal{B}}^{\alpha}(x_2)\mathcal{V}_\mu(x_1)\bar{\chi}_{\mathcal{B}'}^{\alpha'}(0)]|\text{vac}\rangle. \quad (4.18)$$

Here, t_1 is the time when the external electromagnetic field interacts with a quark and t_2 is the time when the final baryon state is annihilated. Γ functions are defined in Eq. (4.8). We choose the baryon interpolating fields, similar to that of the proton, but by replacing the quark fields accordingly as,

$$\begin{aligned}
\chi_{\Sigma_c}(x) &= \varepsilon^{ijk}[\ell^{Ti}(x)C\gamma_5 c^j(x)]\ell^k(x), \\
\chi_{\Xi_{cc}}(x) &= \varepsilon^{ijk}[c^{Ti}(x)C\gamma_5 \ell^j(x)]c^k(x), \\
\chi_{\Omega_c}(x) &= \varepsilon^{ijk}[s^{Ti}(x)C\gamma_5 c^j(x)]s^k(x), \\
\chi_{\Omega_{cc}}(x) &= \varepsilon^{ijk}[c^{Ti}(x)C\gamma_5 s^j(x)]c^k(x),
\end{aligned} \quad (4.19)$$

where $\ell = u$ for the doubly charged $\Xi_{cc}^{++}(ccu)/\Sigma_c^{++}(cuu)$ and $\ell = d$ for the singly charged $\Xi_{cc}^{+}(ccd)/\Sigma_c^{+}(cdd)$ baryons. Here i, j, k denote the color indices and $C = \gamma_4\gamma_2$. Inserting the interpolating fields into Eqs. (4.17) and (4.18) and performing the Wick contractions, we write the correlation functions in terms of propagators. Let us simplify the form of the above interpolating fields by replacing the specific flavours by generic q_1, q_2 and q_3 fields for the ease of discussion,

$$\chi_{\mathcal{B}}(x) = \varepsilon^{ijk}[q_1^{Ti}(x)C\gamma_5 q_2^j(x)]q_3^k(x), \tag{4.20}$$

and write down the two-point and three-point correlation functions in terms of propagators. We rewrite the two-point correlation function in Eq. (4.17) associated with the $\chi_{\mathcal{B}}$ field as,

$$\begin{aligned}
\langle G^{\mathcal{B}\mathcal{B}}(t;\mathbf{p};\Gamma_4)\rangle &= \sum_{\mathbf{x}} e^{-i\mathbf{p}\cdot\mathbf{x}}\Gamma_4^{\alpha\alpha'}\langle \mathrm{vac}|\mathcal{T}[\chi_{\mathcal{B}}^{\alpha}(x)\bar{\chi}_{\mathcal{B}}^{\alpha'}(0)]|\mathrm{vac}\rangle \\
&= \sum_{\mathbf{x}} e^{-i\mathbf{p}\cdot\mathbf{x}}\Gamma_4^{\alpha\alpha'}\varepsilon^{ijk}\varepsilon^{i'j'k'}(\tilde{C})_{\beta\gamma}(\tilde{C})_{\gamma'\beta'}\langle \mathrm{vac}|q_1^i(x)_\alpha q_2^j(x)_\beta q_3^k(x)_\gamma \bar{q}_3^{k'}(0)_{\gamma'}\bar{q}_2^{j'}(0)_{\beta'}\bar{q}_1^{i'}(0)_{\alpha'}|\mathrm{vac}\rangle \\
&= \sum_{\mathbf{x}} e^{-i\mathbf{p}\cdot\mathbf{x}}\varepsilon^{ijk}\varepsilon^{i'j'k'}\Big\{\mathrm{Tr}\Big[\Gamma_4 S_{q_1}^{ii'}(x,0)\underline{S}_{q_2}^{jj'}(x,0)S_{q_3}^{kk'}(x,0)\Big] \\
&\qquad + \mathrm{Tr}\Big[\Gamma_4 S_{q_1}^{ii'}(x,0)\Big]\mathrm{Tr}\Big[\underline{S}_{q_2}^{jj'}(x,0)S_{q_3}^{kk'}(x,0)\Big]\Big\},
\end{aligned} \tag{4.21}$$

where $S_{q_i}^{ii'}(x|0)$ is the quark propagator of the ith quark field, $\alpha^{(\prime)}, \beta^{(\prime)}, \gamma^{(\prime)}$ denote the Dirac indices and we have defined $\tilde{C} \equiv C\gamma_5$ and $\underline{S} \equiv (\tilde{C}S\tilde{C}^{-1})^T$ to make the forms simpler. By inserting the interpolating field into Eq. (4.18), connected pieces of the three-point correlation function becomes,

$$\begin{aligned}
\langle G^{\mathcal{B}\mathcal{V}_\mu\mathcal{B}}(t_2,t_1;\mathbf{p}',\mathbf{p};\Gamma)\rangle &\equiv -i\sum_{\mathbf{x}_2,\mathbf{x}_1} e^{-i\mathbf{p}\cdot\mathbf{x}_2}e^{i\mathbf{q}\cdot\mathbf{x}_1}\Gamma^{\alpha\alpha'}\langle \mathrm{vac}|\mathcal{T}[\chi_{\mathcal{B}}^{\alpha}(x_2)\mathcal{V}_\mu(x_1)\bar{\chi}_{\mathcal{B}'}^{\alpha'}(0)]|\mathrm{vac}\rangle \\
&= \sum_{\mathbf{x}_2,\mathbf{x}_1} e^{-i\mathbf{p}\cdot\mathbf{x}_2}e^{i\mathbf{q}\cdot\mathbf{x}_1}\varepsilon^{ijk}\varepsilon^{i'j'k'}\Big\{\mathrm{Tr}\Big[\Gamma\hat{S}_{q_1}^{ii'}(x_2,x_1,0)\underline{S}_{q_2}^{jj'}(x_2,0)S_{q_3}^{kk'}(x_2,0)\Big] \\
&+ \mathrm{Tr}\Big[\Gamma\hat{S}_{q_1}^{ii'}(x_2,x_1,0)\Big]\mathrm{Tr}\Big[\underline{S}_{q_2}^{jj'}(x_2,0)S_{q_3}^{kk'}(x_2,0)\Big] \\
&+ \mathrm{Tr}\Big[\Gamma S_{q_1}^{ii'}(x_2,0)\underline{S}_{q_2}^{jj'}(x_2,0)\hat{S}_{q_3}^{kk'}(x_2,x_1,0)\Big] \\
&+ \mathrm{Tr}\Big[\Gamma S_{q_1}^{ii'}(x_2,0)\Big]\mathrm{Tr}\Big[\underline{S}_{q_2}^{jj'}(x,0)\hat{S}_{q_3}^{kk'}(x_2,x_1,0)\Big] \\
&+ \mathrm{Tr}\Big[\Gamma S_{q_1}^{ii'}(x_2,0)\hat{\underline{S}}_{q_2}^{jj'}(x_2,x_1,0)S_{q_3}^{kk'}(x_2,0)\Big] \\
&+ \mathrm{Tr}\Big[\Gamma S_{q_1}^{ii'}(x_2,0)\Big]\mathrm{Tr}\Big[\hat{\underline{S}}_{q_2}^{jj'}(x_2,x_1,0)S_{q_3}^{kk'}(x_2,0)\Big]\Big\},
\end{aligned} \tag{4.22}$$

where we have defined the current inserted propagator,

$$\hat{S}_{q_{1-3}}^{ii'}(x_2,x_1,0) \equiv S_{q_{1-3}}^{li'}(x_2,x_1)\mathcal{V}_\mu(x_1)S_{q_{1-3}}^{al'}(x_1,0), \tag{4.23}$$

with $\mathcal{V}_\mu$ denoting the conserved, point-split lattice electromagnetic current,

$$\mathcal{V}_\mu(x) = \frac{1}{2}[\bar{q}(x+\hat{\mu})U_\mu^\dagger(1+\gamma_\mu)q(x) - \bar{q}(x)U_\mu(1-\gamma_\mu)q(x+\hat{\mu})]. \tag{4.24}$$

Once we identify the quark fields, q_{1-3}, according to the Eq. (4.19) we recover the two- and three-point correlation functions of the respective baryons.

Written in terms of the hadron degrees of freedom, two-point correlation function reduces to the Eq. (4.7) we have derived in Sect. 4.1.1.1. Inserting sets of complete states to the left and right of the current, $\mathcal{V}_\mu$, and using the Eqs. (4.5) and (4.13), three-point correlation function takes the form,

$$\begin{aligned}
\langle G^{\mathcal{B}\mathcal{V}_\mu\mathcal{B}}(t_2, t_1; \mathbf{p}', \mathbf{p}; \Gamma)\rangle &= Z_{\mathcal{B}}(\mathbf{p}')\bar{Z}_{\mathcal{B}}(\mathbf{p})\sqrt{\frac{M_{\mathcal{B}}M_{\mathcal{B}}}{E_{\mathcal{B}}(\mathbf{p})E_{\mathcal{B}}(\mathbf{p}')}}e^{-E_{\mathcal{B}}(\mathbf{p}')t_2}e^{[E_{\mathcal{B}}(\mathbf{p}')-E_{\mathcal{B}}(\mathbf{p})]t_1}\Gamma^{\alpha\alpha'} \\
&\times \sum_{s,s'} u^\alpha_{\mathcal{B}}(p', s')\bar{u}^\beta_{\mathcal{B}}(p', s')\left[\gamma_\mu F_{1,\mathcal{B}}(q^2) + i\frac{\sigma_{\mu\nu}q^\nu}{2M_{\mathcal{B}}}F_{2,\mathcal{B}}(q^2)\right]_{\beta\beta'} u^{\beta'}_{\mathcal{B}}(p, s)\bar{u}^{\alpha'}_{\mathcal{B}}(p, s) \\
&= Z_{\mathcal{B}}(\mathbf{p}')\bar{Z}_{\mathcal{B}}(\mathbf{p})\sqrt{\frac{M_{\mathcal{B}}M_{\mathcal{B}}}{E_{\mathcal{B}}(\mathbf{p})E_{\mathcal{B}}(\mathbf{p}')}}e^{-E_{\mathcal{B}}(\mathbf{p}')t_2}e^{[E_{\mathcal{B}}(\mathbf{p}')-E_{\mathcal{B}}(\mathbf{p})]t_1} \\
&\times \mathrm{Tr}\left[\Gamma\frac{\gamma_\mu p'^\mu + M_{\mathcal{B}}}{2M_{\mathcal{B}}}\left(\gamma_\mu F_{1,\mathcal{B}}(q^2) + i\frac{\sigma_{\mu\nu}q^\nu}{2M_{\mathcal{B}}}F_{2,\mathcal{B}}(q^2)\right)\frac{\gamma_\mu p^\mu + M_{\mathcal{B}}}{2M_{\mathcal{B}}}\right]. \quad (4.25)
\end{aligned}$$

Sachs form factors, $G_{E,\mathcal{B}}(q^2)$ and $G_{M,\mathcal{B}}(q^2)$, are isolated when we take the large Euclidean time limit, $t_2 - t_1$ and $t_1 \gg a$, so that the ratio in Eq. (4.16) reduces to the desired form

$$R(t_2, t_1; \mathbf{p}', \mathbf{p}; \Gamma; \mu) \xrightarrow[t_2-t_1\gg a]{t_1\gg a} \Pi(\mathbf{p}', \mathbf{p}; \Gamma; \mu), \quad (4.26)$$

and by choosing appropriate combinations of Lorentz direction μ and projection matrices Γ, we extract the form factors as,

$$\Pi(\mathbf{0}, -\mathbf{q}; \Gamma_4; \mu = 4) = \left[\frac{(E_{\mathcal{B}} + M_{\mathcal{B}})}{2E_{\mathcal{B}}}\right]^{1/2} G_{E,\mathcal{B}}(q^2), \quad (4.27)$$

$$\Pi(\mathbf{0}, -\mathbf{q}; \Gamma_j; \mu = i) = \left[\frac{1}{2E_{\mathcal{B}}(E_{\mathcal{B}} + M_{\mathcal{B}})}\right]^{1/2} \varepsilon_{ijk}\, q_k\, G_{M,\mathcal{B}}(q^2). \quad (4.28)$$

Here, $G_{E,\mathcal{B}}(0)$ gives the electric charge of the baryon. Similarly, the magnetic moment can be obtained from the magnetic form factor $G_{M,\mathcal{B}}$ at zero momentum transfer.

4.1.2.2 Spin-3/2 Elastic Form Factors

For a spin$-3/2 \to$ spin$-3/2$ transition, matrix element of the electromagnetic current is written as,

$$\langle \mathcal{B}_\sigma(p', s')|V^\mu|\mathcal{B}_\tau(p, s)\rangle = \sqrt{\frac{M_{\mathcal{B}}M_{\mathcal{B}}}{E(p)E(p')}}\bar{u}_\sigma(p', s')\mathcal{O}^{\sigma\mu\tau}u_\tau(p, s), \quad (4.29)$$

where $p(s)$ and $p'(s')$ denote the four momentum (spin) of the initial and final states, respectively. $M_{\mathcal{B}}$ is the mass of the baryon, $E(p)$ ($E(p')$) is the energy of the incoming (outgoing) baryon state and $u_\alpha(p, s)$ is the baryon spinor in the Rarita-Schwinger formalism [3]. The tensor, $\mathcal{O}^{\sigma\mu\tau}$, in Eq. (4.29) has been derived in Ref. [4] and here we summarise the relevant parts to our calculation. $\mathcal{O}^{\sigma\mu\tau}$ is given in a Lorentz-covariant form as

$$\mathcal{O}^{\sigma\mu\tau} = -g^{\sigma\tau}\left\{a_1\gamma^\mu + \frac{a_2}{2M_{\mathcal{B}}}P^\mu\right\} - \frac{q^\sigma q^\tau}{(2M_{\mathcal{B}})^2}\left\{c_1\gamma^\mu + \frac{c_2}{2M_{\mathcal{B}}}P^\mu\right\}, \tag{4.30}$$

where $P = p + p'$ and $q = p' - p$. The multipole form factors are then defined in terms of the covariant vertex functions a_1, a_2, c_1 and c_2 as,

$$G_{E0}(q^2) = (1 + \frac{2}{3}\tau)\left\{a_1 + (1+\tau)a_2\right\} - \frac{1}{3}\tau(1+\tau)\left\{c_1 + (1+\tau)c_2\right\}, \tag{4.31}$$

$$G_{E2}(q^2) = \left\{a_1 + (1+\tau)a_2\right\} - \frac{1}{2}(1+\tau)\left\{c_1 + (1+\tau)c_2\right\}, \tag{4.32}$$

$$G_{M1}(q^2) = (1 + \frac{4}{3}\tau)a_1 - \frac{2}{3}\tau(1+\tau)c_1, \tag{4.33}$$

$$G_{M3}(q^2) = a_1 - \frac{1}{2}(1+\tau)c_1, \tag{4.34}$$

with $\tau = -q^2/(2M_B)^2$. These multipole form factors are referred to as electric-charge ($E0$), electric-quadrupole ($E2$), magnetic-dipole ($M1$) and magnetic-octupole ($M3$) multipole form factors.

The two- and three-point correlation functions for spin-3/2 baryons are defined as,

$$\langle G^{\mathcal{B}\mathcal{B}}_{\sigma\tau}(t; \mathbf{p}; \Gamma_4)\rangle \equiv \sum_{\mathbf{x}} e^{-i\mathbf{p}\cdot\mathbf{x}}\Gamma_4^{\alpha\alpha'}\langle \text{vac}|\mathcal{T}[\chi_{\sigma,\alpha}(x)\bar{\chi}_{\tau,\alpha'}(0)]|\text{vac}\rangle, \tag{4.35}$$

$$\begin{aligned}\langle G^{\mathcal{B}V^\mu\mathcal{B}}_{\sigma\tau}(t_2, t_1; \mathbf{p}', \mathbf{p}; \Gamma)\rangle \equiv {} & -i\sum_{\mathbf{x}_2,\mathbf{x}_1} e^{-i\mathbf{p}'\cdot\mathbf{x}_2}e^{i\mathbf{q}\cdot\mathbf{x}_1} \\ & \times \Gamma^{\alpha\alpha'}\langle \text{vac}|\mathcal{T}[\chi_{\sigma,\alpha}(x_2)V^\mu(x_1)\bar{\chi}_{\tau,\alpha'}(0)]|\text{vac}\rangle,\end{aligned} \tag{4.36}$$

where the spin projection matrices Γ and Γ_4 are the ones defined in Eq. (4.8). α, α' denote the Dirac indices and σ, τ are the Lorentz indices of the spin-3/2 interpolating fields. The baryon interpolating fields are chosen, similarly to those of the decuplet Δ^+ baryon as

$$\chi_\mu(x) = \frac{1}{\sqrt{3}}\varepsilon^{ijk}\{2[q_1^{Ti}(x)C\gamma_\mu q_2^j(x)]q_3^k(x) + [q_1^{Ti}(x)C\gamma_\mu q_3^j(x)]q_2^k(x)\}, \tag{4.37}$$

where i, j, k denote the color indices and $C = \gamma_4\gamma_2$. q_1, q_2, q_3 are the quark flavors and chosen as $(q_1, q_2, q_3) = \{(s,s,s), (s,s,c), (s,c,c), (c,c,c)\}$ for Ω, Ω_c^*, Ω_{cc}^* and

Ω_{ccc} baryons, respectively. In principle, the interpolating field in Eq. (4.37) couples to spin-1/2 baryons also and a spin-3/2 projection might be desirable but, it has been shown in Refs. [5, 6] that it has minimal overlap with spin-1/2 states and therefore spin-3/2 projection is not necessary.

Inserting the interpolating field into the two- and three-point correlation functions and performing Wick contractions we derive the expressions of correlation functions in terms of quark propagators as,

$$\langle G^{\mathcal{BB}}_{\sigma\tau}(t;\mathbf{p};\Gamma_4)\rangle \equiv \sum_{\mathbf{x}} e^{-i\mathbf{p}\cdot\mathbf{x}}\Gamma_4^{\alpha\alpha'}\langle \text{vac}|\mathcal{T}[\chi_{\sigma,\alpha}(x)\bar{\chi}_{\tau,\alpha'}(0)]|\text{vac}\rangle \tag{4.38}$$

$$\begin{aligned}
&= \frac{1}{3}\sum_{\mathbf{x}} e^{-i\mathbf{p}\cdot\mathbf{x}}\varepsilon^{ijk}\varepsilon^{i'j'k'}\Big\{4\text{Tr}\Big[\Gamma_4 S^{kk'}_{q_2}(x,0)\gamma_\tau C S^{Tii'}_{q_3}(x,0)C\gamma_\sigma S^{jj'}_{q_1}(x,0)\Big]\\
&+4\text{Tr}\Big[\Gamma_4 S^{kk'}_{q_2}(x,0)\gamma_\tau C S^{Tii'}_{q_1}(x,0)C\gamma_\sigma S^{jj'}_{q_3}(x,0)\Big]\\
&+4\text{Tr}\Big[\Gamma_4 S^{kk'}_{q_3}(x,0)\gamma_\tau C S^{Tii'}_{q_1}(x,0)C\gamma_\sigma S^{jj'}_{q_2}(x,0)\Big]\\
&+2\Big(\text{Tr}\Big[\Gamma_4 S^{kk'}_{q_2}(x,0)\Big]\text{Tr}\Big[\gamma_\tau C S^{Tii'}_{q_3}(x,0)C\gamma_\sigma S^{jj'}_{q_1}(x,0)\Big]\Big)\\
&+2\Big(\text{Tr}\Big[\Gamma_4 S^{kk'}_{q_2}(x,0)\Big]\text{Tr}\Big[\gamma_\tau C S^{Tii'}_{q_1}(x,0)C\gamma_\sigma S^{jj'}_{q_3}(x,0)\Big]\Big)\\
&+2\Big(\text{Tr}\Big[\Gamma_4 S^{kk'}_{q_3}(x,0)\Big]\text{Tr}\Big[\gamma_\tau C S^{Tii'}_{q_1}(x,0)C\gamma_\sigma S^{jj'}_{q_2}(x,0)\Big]\Big)\Big\},
\end{aligned} \tag{4.39}$$

where, similar to the notation in the spin-1/2 case, $S^{ii'}_{q_i}(x,0)$ is the quark propagator of the ith quark field in the interpolator. We derive the connected contributions of the three-point function by inserting the electromagnetic current to each quark propagator in the above expression. There are 18 terms to consider,

$$\begin{aligned}
\langle G^{\mathcal{B}\mathcal{V}^\mu\mathcal{B}}_{\sigma\tau}(t_2,t_1;\mathbf{p}',\mathbf{p};\Gamma)\rangle &\equiv -i\sum_{\mathbf{x}_2,\mathbf{x}_1} e^{-i\mathbf{p}'\cdot\mathbf{x}_2}e^{i\mathbf{q}\cdot\mathbf{x}_1}\times\Gamma^{\alpha\alpha'}\langle \text{vac}|\mathcal{T}[\chi_{\sigma,\alpha}(x_2)\mathcal{V}^\mu(x_1)\bar{\chi}_{\tau,\alpha'}(0)]|\text{vac}\rangle\\
&= \frac{1}{3}\sum_{\mathbf{x}_2} e^{-i\mathbf{p}\cdot\mathbf{x}_2}\varepsilon^{ijk}\varepsilon^{i'j'k'}\Big\{4\text{Tr}\Big[\Gamma S^{kk'}_{q_2}(x,0)\gamma_\tau C S^{Tii'}_{q_3}(x,0)C\gamma_\sigma \hat{S}^{jj'}_{q_1}(x_2,x_1,0)\Big]\\
&+4\text{Tr}\Big[\Gamma S^{kk'}_{q_2}(x,0)\gamma_\tau C \hat{S}^{Tii'}_{q_1}(x_2,x_1,0)C\gamma_\sigma S^{jj'}_{q_3}(x,0)\Big]\\
&+4\text{Tr}\Big[\Gamma S^{kk'}_{q_3}(x,0)\gamma_\tau C \hat{S}^{Tii'}_{q_1}(x_2,x_1,0)C\gamma_\sigma S^{jj'}_{q_2}(x,0)\Big]\\
&+4\text{Tr}\Big[\Gamma \hat{S}^{kk'}_{q_2}(x_2,x_1,0)\gamma_\tau C S^{Tii'}_{q_3}(x,0)C\gamma_\sigma S^{jj'}_{q_1}(x,0)\Big]\\
&+4\text{Tr}\Big[\Gamma \hat{S}^{kk'}_{q_2}(x_2,x_1,0)\gamma_\tau C S^{Tii'}_{q_1}(x,0)C\gamma_\sigma S^{jj'}_{q_3}(x,0)\Big]\\
&+4\text{Tr}\Big[\Gamma S^{kk'}_{q_3}(x,0)\gamma_\tau C S^{Tii'}_{q_1}(x,0)C\gamma_\sigma \hat{S}^{jj'}_{q_2}(x_2,x_1,0)\Big]\\
&+4\text{Tr}\Big[\Gamma S^{kk'}_{q_2}(x,0)\gamma_\tau C \hat{S}^{Tii'}_{q_3}(x_2,x_1,0)C\gamma_\sigma S^{jj'}_{q_1}(x,0)\Big]\\
&+4\text{Tr}\Big[\Gamma S^{kk'}_{q_2}(x,0)\gamma_\tau C S^{Tii'}_{q_1}(x,0)C\gamma_\sigma \hat{S}^{jj'}_{q_3}(x_2,x_1,0)\Big]\\
&+4\text{Tr}\Big[\Gamma \hat{S}^{kk'}_{q_3}(x_2,x_1,0)\gamma_\tau C S^{Tii'}_{q_1}(x,0)C\gamma_\sigma S^{jj'}_{q_2}(x,0)\Big]
\end{aligned}$$

$$
\begin{aligned}
&+2\left(\mathrm{Tr}\left[\Gamma S^{kk'}_{q2}(x,0)\right]\mathrm{Tr}\left[\gamma_\tau C S^{Tii'}_{q3}(x,0)C\gamma_\sigma \hat{S}^{jj'}_{q1}(x_2,x_1,0)\right]\right)\\
&+2\left(\mathrm{Tr}\left[\Gamma S^{kk'}_{q2}(x,0)\right]\mathrm{Tr}\left[\gamma_\tau C \hat{S}^{Tii'}_{q1}(x_2,x_1,0)C\gamma_\sigma S^{jj'}_{q3}(x,0)\right]\right)\\
&+2\left(\mathrm{Tr}\left[\Gamma S^{kk'}_{q3}(x,0)\right]\mathrm{Tr}\left[\gamma_\tau C \hat{S}^{Tii'}_{q1}(x_2,x_1,0)C\gamma_\sigma S^{jj'}_{q2}(x,0)\right]\right)\\
&+2\left(\mathrm{Tr}\left[\Gamma \hat{S}^{kk'}_{q2}(x_2,x_1,0)\right]\mathrm{Tr}\left[\gamma_\tau C S^{Tii'}_{q3}(x,0)C\gamma_\sigma S^{jj'}_{q1}(x,0)\right]\right)\\
&+2\left(\mathrm{Tr}\left[\Gamma \hat{S}^{kk'}_{q2}(x_2,x_1,0)\right]\mathrm{Tr}\left[\gamma_\tau C S^{Tii'}_{q1}(x,0)C\gamma_\sigma S^{jj'}_{q3}(x,0)\right]\right)\\
&+2\left(\mathrm{Tr}\left[\Gamma S^{kk'}_{q3}(x,0)\right]\mathrm{Tr}\left[\gamma_\tau C S^{Tii'}_{q1}(x,0)C\gamma_\sigma \hat{S}^{jj'}_{q2}(x_2,x_1,0)\right]\right)\\
&+2\left(\mathrm{Tr}\left[\Gamma S^{kk'}_{q2}(x,0)\right]\mathrm{Tr}\left[\gamma_\tau C \hat{S}^{Tii'}_{q3}(x_2,x_1,0)C\gamma_\sigma S^{jj'}_{q1}(x,0)\right]\right)\\
&+2\left(\mathrm{Tr}\left[\Gamma S^{kk'}_{q2}(x,0)\right]\mathrm{Tr}\left[\gamma_\tau C S^{Tii'}_{q1}(x,0)C\gamma_\sigma \hat{S}^{jj'}_{q3}(x_2,x_1,0)\right]\right)\\
&+2\left(\mathrm{Tr}\left[\Gamma \hat{S}^{kk'}_{q3}(x_2,x_1,0)\right]\mathrm{Tr}\left[\gamma_\tau C S^{Tii'}_{q1}(x,0)C\gamma_\sigma S^{jj'}_{q2}(x,0)\right]\right)\Big\}, \qquad (4.40)
\end{aligned}
$$

where the current inserted propagator, $\hat{S}_{q_i}(x_2, x_1, 0)$, is defined in Eq. (4.23). We compute each propagator and perform the $\mathbf{x_1}$ summation by the use of *wall-smearing method* as described in Sect. 4.2.5 and calculate the correlation functions as dictated by Eqs. (4.38) and (4.40).

At the hadronic level we derive the expressions for the two- and three-point correlation functions by inserting sets of complete eigenstates $\sum_s |(p,s)\rangle\langle(p,s)|$ into Eqs. (4.35) and (4.36), like we did for the spin-1/2 case,

$$
\langle G^{\mathcal{B}\mathcal{B}}_{\sigma\tau}(t;\mathbf{p};\Gamma_4)\rangle = Z_{\mathcal{B}}(\mathbf{p}')\bar{Z}_{\mathcal{B}}(\mathbf{p})\frac{M_{\mathcal{B}}}{E(\mathbf{p})}e^{-E(\mathbf{p})t}\mathrm{Tr}[\Gamma_4\Lambda_{\sigma\tau}], \qquad (4.41)
$$

$$
\begin{aligned}
\langle G^{\mathcal{B}V^\mu\mathcal{B}}_{\sigma\tau}(t_2,t_1;\mathbf{p}',\mathbf{p};\Gamma)\rangle &= Z_{\mathcal{B}}(\mathbf{p}')\bar{Z}_{\mathcal{B}}(\mathbf{p})\sqrt{\frac{M_{\mathcal{B}}M_{\mathcal{B}}}{E(\mathbf{p})E(\mathbf{p}')}}e^{-E_{\mathcal{B}}(\mathbf{p}')(t_2-t_1)}e^{-E_{\mathcal{B}}(\mathbf{p})t_1}\\
&\times \mathrm{Tr}[\Gamma\Lambda_{\sigma\sigma'}(p')\mathcal{O}^{\sigma'\mu\tau'}\Lambda_{\tau'\tau}(p)], \qquad (4.42)
\end{aligned}
$$

where the trace acts in the Dirac space, the $Z_{\mathcal{B}}(p)$ is the overlap factor of the interpolating field to the corresponding baryon state and $\Lambda_{\sigma\tau}$ is the Rarita-Schwinger spin sum for the spin-3/2 field in Euclidean space, defined as

$$
\begin{aligned}
\sum_s u_\sigma(p,s)\bar{u}_\tau(p,s) &= \frac{-i\gamma\cdot p + M_{\mathcal{B}}}{2M_{\mathcal{B}}}\left[g_{\sigma\tau} - \frac{1}{3}\gamma_\sigma\gamma_\tau + \frac{2p_\sigma p_\tau}{3M^2_{\mathcal{B}}} - i\frac{p_\sigma\gamma_\tau - p_\tau\gamma_\sigma}{3M_{\mathcal{B}}}\right] \qquad (4.43)\\
&\equiv \Lambda_{\sigma\tau}(p).
\end{aligned}
$$

To extract the multipole form factors we consider the large Euclidean time limit, $t_2 - t_1$ and $t_1 \gg a$, of the following ratio of the correlation functions given in Eqs. (4.35) and (4.36),

$$R_{\sigma}{}^{\mu}{}_{\tau}(t_2, t_1; \mathbf{p}', \mathbf{p}; \Gamma) = \left[\frac{\langle G^{\mathcal{B}\mathcal{V}^\mu\mathcal{B}}_{\sigma\tau}(t_2, t_1; \mathbf{p}', \mathbf{p}; \Gamma)\rangle \langle G^{\mathcal{B}\mathcal{V}^\mu\mathcal{B}}_{\sigma\tau}(t_2, t_1; \mathbf{p}, -\mathbf{p}'; \Gamma)\rangle}{\langle G^{\mathcal{B}\mathcal{B}}_{\sigma\tau}(t_2; \mathbf{p}'; \Gamma_4)\rangle \langle G^{\mathcal{B}\mathcal{B}}_{\sigma\tau}(t_2; -\mathbf{p}; \Gamma_4)\rangle} \right]^{1/2}$$

$$\xrightarrow[t_2 - t_1 \gg a]{t_1 \gg a} \left(\frac{E_{\mathcal{B}}(\mathbf{p}) + M_{\mathcal{B}}}{2E_{\mathcal{B}}(\mathbf{p})} \right)^{1/2} \left(\frac{E_{\mathcal{B}}(\mathbf{p}') + M_{\mathcal{B}}}{2E_{\mathcal{B}}(\mathbf{p}')} \right)^{1/2} \Pi_{\sigma}{}^{\mu}{}_{\tau}(\mathbf{p}', \mathbf{p}; \Gamma). \quad (4.44)$$

Note that there is no sum over the repeated indices. The multipole form factors are extracted by using the following combinations of $\Pi_{\sigma}{}^{\mu}{}_{\tau}(\mathbf{p}', \mathbf{p}; \Gamma)$ [6]:

$$G_{E0}(q^2) = \frac{1}{3} \left(\Pi_1{}^4{}_1(\mathbf{q}_i, 0; \Gamma_4) + \Pi_2{}^4{}_2(\mathbf{q}_i, 0; \Gamma_4) + \Pi_3{}^4{}_3(\mathbf{q}_i, 0; \Gamma_4) \right), \quad (4.45)$$

$$G_{E2}(q^2) = 2\frac{M(E+M)}{|\mathbf{q}_i|^2} \left(\Pi_1{}^4{}_1(\mathbf{q}_i, 0; \Gamma_4) + \Pi_2{}^4{}_2(\mathbf{q}_i, 0; \Gamma_4) - 2\Pi_3{}^4{}_3(\mathbf{q}_i, 0; \Gamma_4) \right), \quad (4.46)$$

$$G_{M1}(q^2) = -\frac{3}{5} \frac{E+M}{|\mathbf{q}_1|^2} \left(\Pi_1{}^3{}_1(\mathbf{q}_1, 0; \Gamma_2) + \Pi_2{}^3{}_2(\mathbf{q}_1, 0; \Gamma_2) + \Pi_3{}^3{}_3(\mathbf{q}_1, 0; \Gamma_2) \right), \quad (4.47)$$

$$G_{M3}(q^2) = -4\frac{M(E+M)^2}{|\mathbf{q}_1|^3} \left(\Pi_1{}^3{}_1(\mathbf{q}_1, 0; \Gamma_2) + \Pi_2{}^3{}_2(\mathbf{q}_1, 0; \Gamma_2) - \frac{3}{2}\Pi_3{}^3{}_3(\mathbf{q}_1, 0; \Gamma_2) \right), \quad (4.48)$$

where $i = 1, 2, 3$ and $\mathbf{q}_i$ are the momentum vectors in three spatial directions. In case of the $E2$ form factor, it is possible to exploit the symmetry,

$$\Pi_2{}^4{}_2(\mathbf{q}_i, 0; \Gamma_4) = \Pi_3{}^4{}_3(\mathbf{q}_i, 0; \Gamma_4), \quad (4.49)$$

and define an average,

$$\Pi^4_{avg}(\mathbf{q}_i, 0; \Gamma_4) = \frac{1}{2} \left[\Pi_2{}^4{}_2(\mathbf{q}_i, 0; \Gamma_4) + \Pi_3{}^4{}_3(\mathbf{q}_i, 0; \Gamma_4) \right], \quad (4.50)$$

in order to decrease the statistical noise in G_{E2}. With the above definitions, G_{E2} form factor is rewritten as,

$$G_{E2}(q^2) = 2\frac{M(E+M)}{|\mathbf{q}_i|^2} \left(\Pi_1{}^4{}_1(\mathbf{q}_i, 0; \Gamma_4) - \Pi^4_{avg}(\mathbf{q}_i, 0; \Gamma_4) \right). \quad (4.51)$$

We consider an average over momentum directions for both $E0$ and $E2$ form factors. In case of the $M1$ form factor, we make a redefinition to utilise all possible index combinations in order to improve the signal. Sum of all correlation-function ratios for $M1$ is written as,

$$G_{M1}(q^2) = -\frac{3}{5} \frac{(E+M)}{|\mathbf{q}|^2} \frac{1}{6} \sum_{\substack{i,j,k=1 \\ i \neq j \neq k}}^{3} \left[\Pi_i{}^j{}_i(\mathbf{q}_i, 0; \Gamma_k) + \Pi_i{}^j{}_i(\mathbf{q}_k, 0; \Gamma_i) + \Pi_i{}^i{}_i(\mathbf{q}_j, 0; \Gamma_k) \right]. \quad (4.52)$$

Compared to the dominant form factors $E0$ and $M1$ we have observed that the data for the $E2$ and $M3$ form factor is much noisier. It turns out that with the limited number of gauge configurations in ensemble N5, the data for $M3$ moments are too noisy to allow a statistically significant value. Thus, we omit the $M3$ form factor in this work and extract only the $E0$, $M1$ and $E2$ form factors for the lowest allowed lattice momentum transfer.

It is possible that the higher order form factors in the expansion interfere with the leading and sub-leading form factors that we consider. Although a dedicated study would be needed to have a strong conclusion we note that the agreement between the results obtained from different lattice formulations of Refs. [6, 7] suggest that the interference effects are minimal.

4.1.3 Data Analysis

We briefly summarise our statistical analysis method rather than going into detail since it can be found in any statistics textbook. For instance, *Statistics for nuclear and particle physicists* by Louis Lyons [8] is a good pedagogical book with an emphasis on nuclear and particle physics.

The quantities that we extract in this work, be it either the mass of the hadron or a form factor value, are all fit results to data points that are varying with respect to time t or four-momentum Q^2 or m_π^2. Let us call them as a generic parameter z and the set of corresponding data points as $\{z\}$. We use least squares regression analysis in order to estimate the observables. Considering an uncorrelated fit, we minimize the function,

$$\mathcal{F}(z) = \sum_{\{z\}} \left(\frac{f(z) - f'(z)}{\sigma(z)} \right)^2, \tag{4.53}$$

with respect to the parameters of the function $f'(z)$, where $f(z)$ is the value of the data set (i.e. correlation function or the ratio) at point z and $\sigma(z)$ is the standard deviation associated with that point. The function $f'(z)$ might have a linear or non-linear dependence to its free parameters depending on the observable we wish to extract and its form is mentioned in each case we perform a regression analysis in the respective sections.

We test the goodness of the fits by the χ^2-method. Minimised function in Eq. (4.53) is mapped to a χ^2 distribution when the data follows a Gaussian distribution. Depending on the number of degrees of freedom, a χ^2 distribution has a varying form with the property $\overline{\chi}^2 = \#$ d.o.f.. Goodness of fit is deduced by comparing the value of the minimised function, $\mathcal{F}(z)$, to the χ^2 value of the corresponding distribution. If $\chi^2 \geq \mathcal{F}(z)$, then we have a reliable fit and the parameters that we have extracted can be trusted. χ^2 values can be looked-up from relevant tables or we can simplify the process by normalising the $\mathcal{F}(z)$ by the number of degrees of freedom and comparing it to 1.0. Large deviations in $\mathcal{F}(z)/\#$ d.o.f. compared to 1.0 or having exactly

1.0 is an indication of something suspicious going-on with the fits and should be investigated further, or, the data set can not be modelled by the function $f'(z)$ and should be discarded.

We estimate all statistical errors by a single-elimination jackknife method [9, 10]. An important issue to note about this method is that the resampling of the data changes the definition of the standard deviation. Originally, standard deviation of a sample is calculated by the formula,

$$\sigma = \sqrt{\frac{1}{N}\sum_{n=1}^{N}(x_n - \overline{x})^2}, \quad \overline{x} = \frac{1}{N}\sum_{n=1}^{N} x_n, \tag{4.54}$$

where N is the sample size and x_n is the nth value of the data set. Definition changes to,

$$\sigma = \sqrt{\frac{N-1}{N}\sum_{n=1}^{N}(x_n - \overline{x})^2}, \tag{4.55}$$

for a resampled data set by a single-elimination jackknife method.

In principle, there may be correlations among the data points or samples which should be accounted by a correlated fit. However, a correlated fit might be rather unstable as pointed out in Refs. [11, 12] and unnecessary if the data sets are small. Regarding that conclusion and personal trial and error, we analyse the data via uncorrelated fits.

4.2 Simulation Setup

In the following sections we go through the details of the gauge configurations and methods that we employ in order to compute the matrix elements mentioned in Sect. 4.1.2.

4.2.1 *Gauge Configurations*

We have run our lattice simulations on $32^3 \times 64$ lattices with 2 + 1 flavors of dynamical quarks using the gauge configurations generated by the PACS-CS collaboration [13] with the nonperturbatively $\mathcal{O}(a)$-improved Wilson quark action (Clover action) and the Iwasaki gauge action. Properties of the lattices are given in Table 4.1. A lattice spacing of $a = 0.0907(13)$ fm, corresponds to $a^{-1} = 2.176(31)$ GeV, which we use in order to convert the lattice results into physical units. We utilize five different ensembles with hopping parameters for the sea and the u,d valence quarks, $\kappa_{sea}, \kappa_{val}^{u,d} = 0.13700, 0.13727, 0.13754, 0.13770$ and 0.13781, which correspond to

Table 4.1 Details of the gauge configurations used in this work. We list the number of flavors (N_f), the lattice spacing (a), the lattice size (L), inverse gauge coupling (β), clover coefficient (c_{SW}) and number of gauge configurations employed (N_{gc}). Corresponding pion masses (m_π) as calculated by PACS-CS Collaboration (Table 3 of Ref. [13]) are also given as a reference

ID	$N_s^3 \times N_t$	N_f	a [fm]	L [fm]	β	$c_{SW}^{u/d,s}$	κ_{val}^s	c_{SW}^c	κ_{val}^c	$\kappa_{val}^{u,d}$	m_π [MeV]
N1								1.52617		0.13700	702(11)
N2								1.52493		0.13727	570(10)
N3	$32^3 \times 64$	$2+1$	0.0907(13)	2.90	1.90	1.715	0.13640	1.52381	0.1246	0.13754	411(8)
N4								1.52326		0.13770	296(7)
N5								1.52264		0.13781	156(9)

pion masses of ~700, 570, 410, 300 and 156 MeV, respectively. We have assigned ID numbers to the ensembles in order to ease further referrals. Strange quark hopping parameter is fixed to κ_{sea}, $\kappa_{val}^s = 0.13640$ and we have determined the charm quark parameter—as outlined in Sect. 4.2.2—to be $\kappa_{val}^c = 0.1246$.

4.2.2 κ_c tuning

It is necessary to compute the quark propagators when calculating the two- and three-point correlation functions which are needed to extract the matrix elements. Considering the light, u/d and s, quarks we use the PACS-CS determined values of the hopping parameters on each ensemble as given in Table 4.1. However, we have to determine the charm quark hopping parameter to compute the charm quark propagators.

We have employed the Clover action to compute the charm quark propagators in order to be consistent with the dynamical quarks. However, one should keep in mind that the Clover action has $\mathcal{O}(a\, m_q)$ discretisation errors, whose effects must be accounted for in case of the charm quark since its larger mass would enhance the discretisation error. As a first precaution we apply the Fermilab method [14] in the form employed by the Fermilab Lattice and MILC Collaborations [15, 16]. A similar procedure has been used to study charmonium, heavy-light meson resonances and their scattering with pion and kaon [17–19]. In the Fermilab method's simplest application one sets the Clover coefficients $c_E = c_B = c_{SW}$ to the tadpole-improved value $1/u_0^3$, where u_0 is the average link. We follow the approach used in Ref. [17] to estimate the u_0 as the fourth root of the average plaquette and determine the charm-quark hopping parameter κ_{val}^c nonperturbatively. We have tuned the κ_c with respect to the spin-averaged static masses of charmonium and open-charmed mesons. As a first crude estimate of κ_{val}^c one can use the vector Ward identity,

$$a\, m_q = \frac{1}{\kappa} - \frac{1}{\kappa_{crit.}}, \tag{4.56}$$

Table 4.2 Charmed meson masses extracted on ensembles N1-N4, as well as linear and quadratic fits to the physical quark-mass point. We also quote the experimental and PACS-CS results for comparison. All values are given in GeV

ID	m_{η_c}	$m_{J/\Psi}$	m_D	m_{D^*}	m_{D_s}	$m_{D_s^*}$
N1	3.019(3)	3.116(5)	2.027(5)	2.180(10)	2.075(5)	2.220(9)
N2	3.006(3)	3.097(4)	1.982(5)	2.112(12)	2.052(4)	2.179(8)
N3	2.992(3)	3.079(4)	1.934(8)	2.077(16)	2.033(5)	2.155(8)
N4	2.984(2)	3.071(3)	1.915(9)	2.045(16)	2.028(4)	2.156(7)
Lin. Fit	2.979(2)	3.063(3)	1.895(6)	2.021(13)	2.018(4)	2.138(7)
Quad. Fit	2.977(4)	3.064(5)	1.893(9)	2.035(22)	2.022(7)	2.156(13)
Exp.	2.980	3.097	1.865	2.007	1.968	2.112
PACS-CS [21]	2.986(1)(13)	3.094(1)(14)	1.871(10)(8)	1.994(11)(9)	1.958(2)(9)	2.095(3)(10)

where m_q is the mass of the quark, a the lattice spacing, κ the hopping parameter of the quark q and $\kappa_{crit.}$ is the critical kappa value where the $m_q \to 0$. Before using this formula however we should determine the $\kappa_{crit.}$ value. Considering the fit ansatz given in the Eq. 40 of Ref. [20] and simplifying it for our case, we can relate the pseudoscalar meson mass, m_π, to the $\kappa_{crit.}$ as

$$m_{PS}^2(\kappa_{val}) = bm_{val}^{VMI} + c\left(m_{val}^{VMI}\right)^2 . \tag{4.57}$$

A fit to the m_π values of the ensembles N1-N4 yields the critical hopping parameter, $\kappa_{crit.} = 0.13787$, which we input to Eq. (4.56). A back-of-the-envelope calculation using the experimental charm quark mass, $m_c = 1.27$ GeV, and the $\kappa_{crit.}$ returns an estimate for the charm quark hopping parameter as, $\kappa_{val}^c = 0.12$. Starting from this simple estimate we have calculated the static masses of the lowest-lying psuedoscalar and vector charmonium states and open-charmed mesons with several κ_{val}^c values on ensembles N1-N4 and determined the charm quark hopping parameter as $\kappa_{val}^c = 0.1246$. We show the effective mass plots of the mesons calculated on all ensembles in Fig. 4.1. All the mass values we have extracted are compiled into Table 4.2 along with their extrapolated values to the physical light-quark mass point. A comparison of the extrapolated and experimental results is illustrated in Fig. 4.2. We discuss the details of the extrapolations in Sect. 5.1.

4.2.3 *Source Smearing*

In this work, we are interested in the ground-state observables. A common practice to improve the ground state dominance of a given correlation function is to *smear* the interpolating fields that are used to create/annihilate the hadrons. Interpolating fields are transformed into propagators by the Wick contraction procedure, so, the components of the actual computations are the propagators rather than the interpolating fields. A propagator on the lattice is computed by inverting the Dirac operator given in Eq. (3.47). Simply put, we need to solve the system of equations

$$DG = S, \tag{4.58}$$

where D is the Dirac operator, S is a source vector and G is the propagator we wish to compute. We are not going to discuss how to solve this system, for it is a research area of its own and many lattice field theory textbooks or books on numerical methods cover the issue, but focus on the source. We basically need to choose a starting point for the propagator and place a quark source onto that point. Let us say we positioned the source on the point x and the simplest form we can assign is a δ-function,

$$S(n) = \delta_{n,m}, \tag{4.59}$$

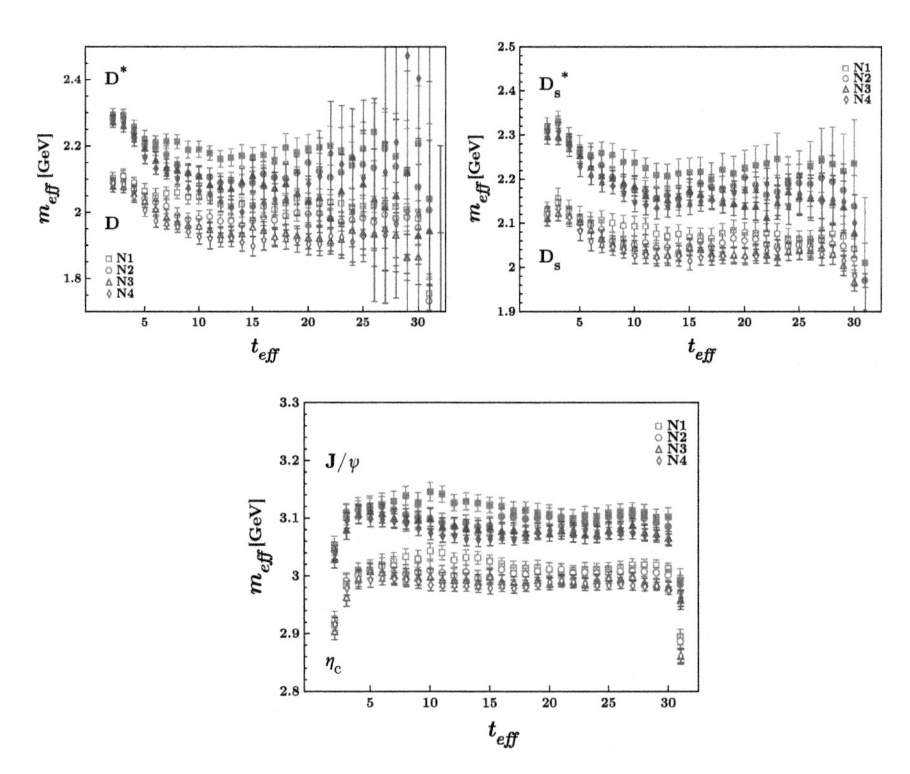

Fig. 4.1 Effective mass plots of the open-charm mesons and charmonium states calculated on ensembles N1-N4 with $\kappa_c = 0.1246$. Open symbols indicate psuedoscalar ($J^P = 0^-$) states while filled symbols are for vector ($J^P = 1^+$) states. Fit regions are chosen as $[t^i_{eff}, t^f_{eff}] = [10, 29]$ for all mesons on all ensembles

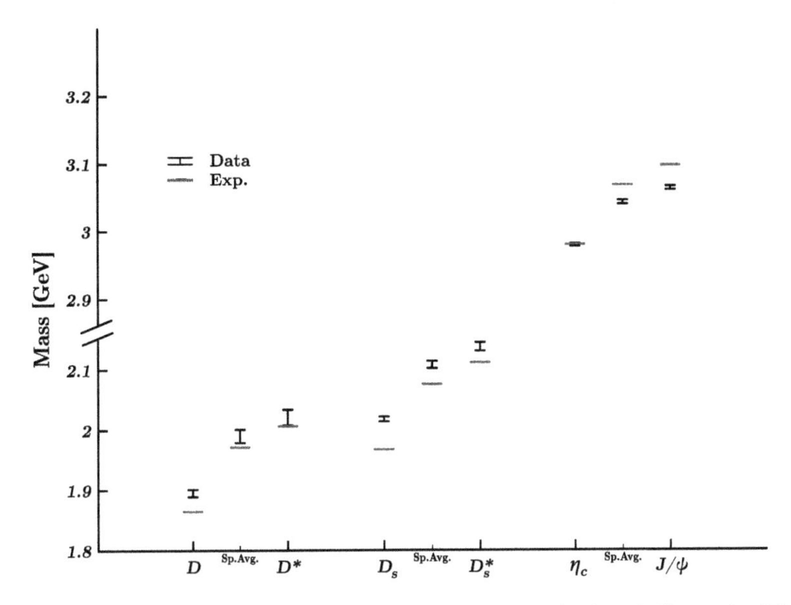

Fig. 4.2 Comparison of the masses to the experimental values. Sp.Avg. indicates the 1 S spin averaged values, i.e. $m_{1S} = (m_{PS} + 3\,m_V)/4$

where we have suppressed the color and Dirac indices for simplicity. This choice of a *point* quark source for the propagator is perfectly acceptable and usually used in case of the sink-operators for connecting propagators but in practice a source operator smeared by a Gaussian form has better ground-state dominance. Gaussian smearing effectively extends the source to neighbouring spatial lattice sites and mimics the extended nature of the hadron since it is not a point-like object. A gauge-invariant method to smear the source is called the Jacobi smearing in which we act on the point-source with a smearing operator, say M, and obtain the smeared source:

$$S'(n) = MS(n) \quad , \quad M = \sum_{i=0}^{N} \alpha^i H^i, \tag{4.60}$$

where α is a positive-real parameter, N is the number of smearing steps and H is the spatial part of the Dirac operator,

$$H(n,m) = \sum_{\mu=1}^{3} \left(U_\mu(n)\delta_{n+\hat{\mu},m} + U_\mu^\dagger(n-\mu)\delta_{n-\hat{\mu},m} \right), \tag{4.61}$$

By adjusting the smearing parameters α and N we can control the width of the source and tune the signal according to our needs. Intuitively parameters are adjusted to reproduce a root mean square radius of $\sim$1 fm.

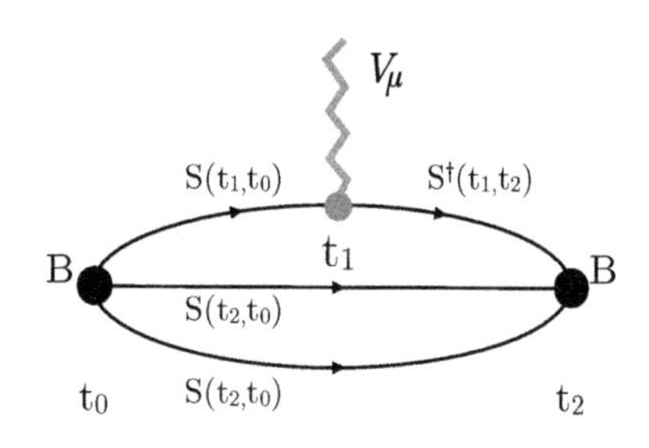

Fig. 4.3 Sketch of a 3-point function. **B** is the baryon, t_0, t_1 and t_2 are source, current insertion and sink time-slices respectively

4.2.4 Fixing the Sink

Form factor calculations involve computing the three-point correlation functions as given in Eqs. (4.18) and (4.36), which encode the interaction with the hadron in question. A closer look to the definitions reveals that one should choose a coordinate for the annihilation operator—or a *sink point*—for the hadron on the discrete space-time. Same is true for the creation operator however its coordinate can always be shifted to the origin using the translational invariance principle. Considering that we have chosen a sink point, we may visualise the connected piece of a three-point correlation function as in Fig. 4.3, where the hadron (denoted by its quark propagator lines) propagates from its source, t_0, to sink point, t_2, while it interacts with an external current at varying t_1.

Let us for a moment recall the two-point correlation function written in terms of the hadron degrees of freedom,

$$\langle O(t)O(0)\rangle_{t\to\infty} = Ae^{-Et}(1 + \mathcal{O}(e^{-\Delta Et})), \tag{4.62}$$

where A is a constant, E is the energy of the hadron and ΔE is the energy difference to the first excited state. Given that the time is large enough, we know that the quantities that we extract belong to the ground state of the hadron. It is an immediate concern however if the number of time slices that the hadron can propagate is limited. In that case the second term in the parentheses might have significant effects and the ground-state observable we are interested in would actually have excited state contributions. Such is a caveat one should be aware of when fixing the sink point to calculate the three-point correlation functions. Time seperation between the source and the sink point should be chosen to be large enough to ensure that the excited states die off so that their effects are negligible.

In our simulations, the source-sink time separation is fixed to $\sim$1.09 fm ($t_2 = 12a$). Statistics limit the upper value of t_2 since the statistical errors grow rapidly as we increase the separation. Therefore, we must choose the smallest possible separation ensuring that the excited-state contaminations are avoided. As for the nucleon axial and electromagnetic form factors, a separation of $\sim$1 fm has been found to be sufficient [2, 22]. A similar conclusion has been made for the Ω^- baryon's electromagnetic form factors [7]. We make further checks in Sects. 5.3 and 5.4 to make sure that the excited state contamination is indeed under control with a choice of $\sim$1.09 fm source-sink seperation.

4.2.5 Sink Smearing

Let us turn our attention to Fig. 4.3 and take a closer look to the propagator connecting the t_1 time slice to t_2. Although all other propagator lines originate from t_0, which can be computed by a single inversion for one quark type, we need an extra inversion to calculate the $S(t_1, t_2)$ propagator and find a way to connect it to the other propagator lines. Supposing we have computed $S(t_1, t_2)$, one way to connect it to other propagators is using the *sequential source method* [23]. This method is widely used in the community but we have opted-out from employing this method because it would have required us to consume more resources due to the following reason. In a nutshell, there are two possibilities to compute the three-point function,

$$\langle G^{\mathcal{B}\mathcal{V}_\mu\mathcal{B}}(t_2, t_1; \mathbf{p}', \mathbf{p}; \Gamma)\rangle \equiv -i\sum_{\mathbf{x}_2,\mathbf{x}_1} e^{-i\mathbf{p}\cdot\mathbf{x}_2} e^{i\mathbf{q}\cdot\mathbf{x}_1} \Gamma^{\alpha\alpha'} \times \langle \text{vac}|\mathcal{T}[\chi^{\alpha}_{\mathcal{B}}(x_2)\mathcal{V}_\mu(x_1)\bar{\chi}^{\alpha'}_{\mathcal{B}'}(0)]|\text{vac}\rangle.$$

(i) We can assign a source to simulate the external current which is associated with the $\mathbf{x}_1$ sum and essentially fix the current operator and the spatial momentum transfer $\mathbf{q}$ or (ii) create a source object containing the two non-interacting propagator lines, which is associated with the $\mathbf{x}_2$ sum, which fixes the sink particle but leaving the current operator free. While the first option is plausible to investigate a specific form factor of various hadrons at fixed momentum, second choice is better suited to study the momentum dependence of different form factors of a specific hadron. What we wish to achieve in this work however, calls for a combination of the two options. Insisting on employing sequential sources would have forced us to invert extra propagators for each different baryon or momentum transfer that we consider, which would increase our consumption of valuable and limited resources although a workaround is available.

An approach that does not require to fix the current operator or the sink hadron is the *wall-smearing method*, where we evaluate the $\mathbf{x}_2$ summation over the spatial sites at the sink time point, *before* computing the propagator so that the propagator (instead of the hadron state) is projected on to definite momentum. In practice, this corresponds to choosing a *wall-smeared* source on the t_2 time slice,

$$S(t_2) = \sum_{x=0}^{N_S-1} \delta_{n,m}\delta_{t,t_2},$$

to compute the $S(t_1, t_2)$ propagator. $S^\dagger(t_1, t_2)$ shown in Fig. 4.3 is defined as $S^\dagger(t_1, t_2) = \gamma_5 S(t_1, t_2)\gamma_5$. $S(t_2, t_0)$ is computed with Gaussian smeared sources and their sinks are wall smeared to match the $S(t_1, t_2)$ propagator. $S(t_1, t_0)$ is already a part of the $S(t_{any}, t_0)$ propagator hence there is no need for an extra inversion. Sinks of the $S(t_1, t_0)$ and $S(t_1, t_2)$ are point smeared. In this way we avoid the necessity of extra inversions for each momentum or hadron state since we are free to choose the current operator and momentum and the sink hadron as long as we have the quark propagators corresponding to the constituents of the hadron we investigate.

4.2.6 Statistical Improvements

Now that we have the freedom to choose the momentum, we can use different momentum components having an equivalent q^2 value in order to increase the statistics. We can also calculate all the spin projections and states corresponding to different Lorentz indices and average over them accordingly. We have different approaches for the spin-1/2 and spin-3/2 cases, so let us itemise the setups for clarity:

- **Spin-1/2**: We insert momentum through the current up to nine units: $(|q_x|,|q_y|,|q_z|)$ $= (0, 0, 0)$, $(1, 0, 0)$, $(1, 1, 0)$, $(1, 1, 1)$, $(2, 0, 0)$, $(2, 1, 0)$, $(2, 1, 1)$, $(2, 2, 0)$, $(2, 2, 1)$ and average over all the possible combinations of the positive and negative momenta in case of the electric form factor. For the magnetic form factor we average over all equivalent combinations of spin projection, Lorentz component and momentum indices.
- **Spin-3/2**: We make our simulations with the lowest allowed lattice momentum transfer $q = 2\pi/(N_s a)$, where N_s is the spatial dimension of the lattice and a is the lattice spacing. This corresponds to three-momentum squared value of $\mathbf{q}^2 = 0.183$ GeV^2. We insert all possible momentum components, namely $(|q_x|, |q_y|, |q_z|)$ $= (-1, 0, 0)$, $(0, -1, 0)$, $(0, 0, -1)$, $(1, 0, 0)$, $(0, 1, 0)$, $(0, 0, 1)$. We also consider vector-current and spin projections along all spatial directions and take into account all Lorentz components of the Rarita-Schwinger field.

In addition to averaging over equivalent combinations, we can exploit the translational invariance principle and shift the source-sink pair on the temporal direction and increase the number of measurements. We have a seperation of $12a$ between the source and the sink point so we shift the source-sink pairs by $12a$ units and increase the number of measurements as necessary.

References

1. S. Aoki, M. Fukugita, S. Hashimoto, Y. Iwasaki, K. Kanaya, Y. Kuramashi, H. Mino, M. Okawa, A. Ukawa, T. Yoshie, Analysis of hadron propagators with 1000 configurations on a $24 * *3x64$ lattice at beta = 6. Nucl. Phys. Proc. Suppl. **47**, 354–357 (1996). https://doi.org/10.1016/0920-5632(96)00072-2
2. C. Alexandrou, M. Brinet, J. Carbonell, M. Constantinou, P.A. Harraud, et al., Nucleon electromagnetic form factors in twisted mass lattice QCD. Phys. Rev., D83: 0 094502 (2011a). https://doi.org/10.1103/PhysRevD.83.094502
3. R. William, J. Schwinger, On a theory of particles with half-integral spin. Phys. Rev. **60**, 61–61 (1941). https://doi.org/10.1103/PhysRev.60.61
4. S. Nozawa, D.B. Leinweber, Electromagnetic form-factors of spin 3/2 baryons. Phys. Rev. D **42**, 3567–3571 (1990). https://doi.org/10.1103/PhysRevD.42.3567

5. C. Alexandrou, V. Drach, K. Jansen, C. Kallidonis, G. Koutsou, Baryon spectrum with $N_f = 2+1+1$ twisted mass fermions. Phys. Rev. **D90 0** (7): 0 074501 (2014).https://doi.org/10.1103/PhysRevD.90.074501
6. S. Boinepalli, D.B. Leinweber, P.J. Moran, A.G. Williams, J.M. Zanotti, J.B. Zhang, Electromagnetic structure of decuplet baryons towards the chiral regime. Phys. Rev. D **80**: 0 054505 (2009). https://doi.org/10.1103/PhysRevD.80.054505
7. C. Alexandrou, T. Korzec, G. Koutsou, J.W. Negele, and Y. Proestos. The Electromagnetic form factors of the Ω^- in lattice QCD. *Phys.Rev.* D82: 0 034504, 2010. https://doi.org/10.1103/PhysRevD.82.034504
8. L. Lyons, *Statistics for nuclear and particle physicists*, (Cambridge University Press, 1986). http://www.cambridge.org/uk/catalogue/catalogue.asp?isbn=0521255406
9. J.W. Tukey, Abstracts of papers. Ann. Math. Statistics **29 0**(2): 0 614–623 (1958). ISSN 00034851. http://www.jstor.org/stable/2237363
10. M.H. Quenouille, Approximate tests of correlation in time-series. J. R. Statistical Soc. Series B (Methodological), **11 0** (1): 0 68–84 (1949). ISSN 00359246. http://www.jstor.org/stable/2983696
11. C. Michael, Fitting correlated data. Phys. Rev. D **49**, 2616–2619 (1994). http://link.aps.org/doi/10.1103/PhysRevD.49.2616
12. C. Michael, A. McKerrell, Fitting correlated hadron mass spectrum data. Phys. Rev. D **51**, 3745–3750 (1995). https://doi.org/10.1103/PhysRevD.51.3745
13. S. Aoki, K.-I. Ishikawa, N. Ishizuka, T. Izubuchi, D. Kadoh, K. Kanaya, Y. Kuramashi, Y. Namekawa, M. Okawa, Y. Taniguchi, A. Ukawa, N. Ukita, T. Yoshie, 2+1 Flavor Lattice QCD toward the Physical Point. Phys. Rev. **D79**: 0 034503 (2009). https://doi.org/10.1103/PhysRevD.79.034503
14. Aida X. El-Khadra, Andreas S. Kronfeld, Paul B. Mackenzie, Massive fermions in lattice gauge theory. Phys. Rev. D **55**, 3933–3957 (1997). https://doi.org/10.1103/PhysRevD.55.3933
15. T. Burch, C. DeTar, M. Di Pierro, A.X. El-Khadra, E.D. Freeland, et al., Quarkonium mass splittings in three-flavor lattice QCD. Phys.Rev. **D81**: 0 034508 (2010). https://doi.org/10.1103/PhysRevD.81.034508
16. C. Bernard, C. DeTar, M. DiPierro, A.X. El-Khadra, R.T. Evans, E.D. Freeland, E. Gamiz, S. Gottlieb, U.M. Heller, J.E. Hetrick, A.S. Kronfeld, J. Laiho, L. Levkova, P. B. Mackenzie, J.N. Simone, R. Sugar, D. Toussaint, R.S./ VandeWater, Tuning fermilab heavy quarks in 2+1 flavor lattice QCD with application to hyperfine splittings. Phys.Rev. **D83**: 0 034503 (2011). https://doi.org/10.1103/PhysRevD.83.034503
17. D. Mohler, R.M. Woloshyn, D and D_s meson spectroscopy. Phys.Rev. **D84**: 0 054505 (2011). https://doi.org/10.1103/PhysRevD.84.054505
18. D. Mohler, S. Prelovsek, R.M. Woloshyn, $D\pi$ scattering and D meson resonances from lattice QCD. Phys.Rev. **D87 0**(3): 0 034501 (2013a). https://doi.org/10.1103/PhysRevD.87.034501
19. D. Mohler, C.B. Lang, L. Leskovec, S. Prelovsek, R.M. Woloshyn, $D^*_{s0}(2317)$ meson and D-meson-kaon scattering from lattice QCD. Phys. Rev. Lett. **111 0**(22): 0 222001 (2013b). https://doi.org/10.1103/PhysRevLett.111.222001
20. A. Ali Khan, S. Aoki, G. Boyd, R. Burkhalter, S. Ejiri, M. Fukugita, S. Hashimoto, N. Ishizuka, Y. Iwasaki, K. Kanaya, T. Kaneko, Y. Kuramashi, T. Manke, K. Nagai, M. Okawa, H.P. Shanahan, A. Ukawa, T. Yoshié, Light hadron spectroscopy with two flavors of dynamical quarks on the lattice. Phys. Rev. D **65 0**(5): 0 054505 (2002). https://doi.org/10.1103/hysRevD.65.054505
21. Y. Namekawa et al., Charm quark system at the physical point of 2+1 flavor lattice QCD. Phys.Rev. **D84**: 0 074505 (2011). https://doi.org/10.1103/PhysRevD.84.074505
22. C. Alexandrou, M. Brinet, J. Carbonell, M. Constantinou, P.A. Harraud, P. Guichon, K. Jansen, T. Korzec, M. Papinutto, Axial nucleon form factors from lattice QCD. Phys. Rev. **D83**: 0 045010 (2011b). https://doi.org/10.1103/PhysRevD.83.045010
23. B. Bunk, K.H. Mutter, K. Schilling, Lattice gauge theory. A challenge in large scale computing, in *Proceedings, NATO workshop*, ed. by F.R. Wuppertal, Germany, November 5–7 (1985). NATO Sci. Ser. B **140**: 0 pp. 1–334 (1986)

Chapter 5
Results and Discussions

Abstract Having set up the necessary formalism to investigate the structural properties of the baryons in the previous chapter, we now apply them to various spin-1/2 and spin-3/2 charmed baryon systems. We first extract the ground-state masses of the baryons in question, which play a role in calculating the form factors apart from their spectroscopic interest. Some further technical details of data analysis crucial for form factor analysis follows along with the interpretation to compute the electromagnetic observables. Thereafter, we present a wealth of results on the form factors, observables such as charge radii and magnetic moments and discuss their implications on the charmed baryons in question. Systematic effects mentioned and analysed along the chapter are collected and summarized at the end of the chapter for clarity. This chapter is essentially based on the following works: K.U. Can, G. Erkol, B. Isildak, M. Oka, T.T. Takahashi, Physics Letters B726 (2013) 703-709, K.U. Can, G. Erkol, B. Isildak, M. Oka, T.T. Takahashi, Journal of High Energy Physics (JHEP) 05, 125 (2014) and K.U. Can, G. Erkol, M. Oka, T.T. Takahashi, Physical Review D 92, 114515 (2015).

Keywords Low-lying charmed baryons
Spin-1/2 and spin-3/2 electromagnetic form factors · Chiral extrapolation
Systematic effects/errors · Charge radii and magnetic moments

Below, we give numerical results and discuss the findings of this work. In Sect. 5.1 we present the baryon masses and discuss possible systematic uncertainties that might manifest themselves in spectroscopy analysis and how to account for them. We have a brief discussion on how we evaluate the lattice data and extract the observables in Sect. 5.2. Sections 5.3 and 5.4 are dedicated to the electromagnetic properties of the spin-1/2 and spin-3/2 baryons. Sources of systematic errors are compiled in Sect. 5.5.

K. U. Can, *Electromagnetic Form Factors of Charmed Baryons in Lattice QCD*,
Springer Theses, https://doi.org/10.1007/978-981-10-8995-4_5

5.1 Baryon Masses

Hadron spectroscopy by itself is a hot topic among the lattice QCD practitioners since it is essential to understand the hadron zoo to understand the strong interactions and being able to calculate the hadron masses from first principles is a powerful alternative to the model calculations or experiments. A recent review on the lattice QCD effort can be found in Ref. [1]. With the advent of progress in experimental efforts and upcoming facilities, spectroscopy of charmed baryons have caught the attention of several groups who have performed systematic analysis of charmed baryon spectroscopy in detail [2–5].

In this work, rather than a precise spectroscopy calculation, we are concerned with the electromagnetic structure of the charmed baryons. From such a point of view,

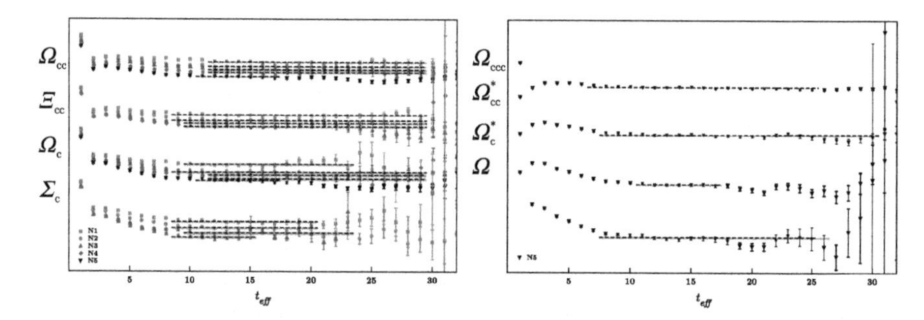

Fig. 5.1 Effective mass plots of the spin-1/2 and spin-3/2 baryons. Note that we have shifted the spin-1/2 data points for the ease of viewing

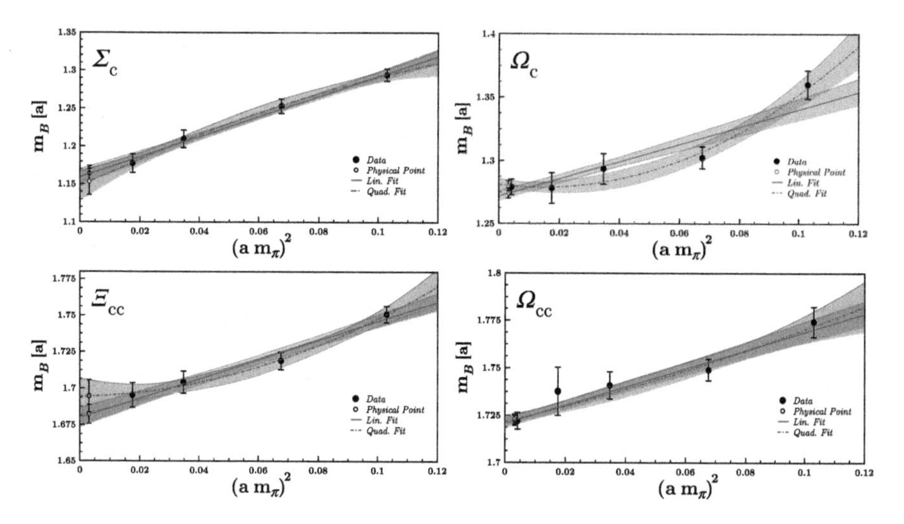

Fig. 5.2 Extrapolations to the physical-mass point using functions given in Eq. (5.1) for the masses of Σ_c, Ξ_{cc}, Ω_c and Ω_{cc} baryons

Table 5.1 The charmed baryon masses extracted on ensembles N1–N5, as well as linear and quadratic fits to the physical quark-mass point. We also quote the experimental and those obtained by PACS-CS [2] (at the physical point, except Ω which is at $m_\pi = 156$ MeV [6]). Results of ETMC [3], Briceno et al. [4] and Brown et al. [5] are also included. Errors in parentheses are statistical for our values while we quote the combined statistical and systematical errors for other collaborations. All results are given in GeV

ID	$\Sigma_c\,(\frac{1}{2}^+)$	$\Omega_c\,(\frac{1}{2}^+)$	Ξ_{cc}	Ω_{cc}	$\Omega\,(\frac{3}{2}^+)$	Ω_c^*	Ω_{cc}^*	Ω_{ccc}
N1	2.841(18)	2.959(24)	3.810(12)	3.861(17)	–	–	–	–
N2	2.753(19)	2.834(19)	3.740(13)	3.806(12)	–	–	–	–
N3	2.647(19)	2.815(26)	3.708(16)	3.788(16)	–	–	–	–
N4	2.584(28)	2.781(26)	3.689(18)	3.781(28)	–	–	–	–
N5	2.486(47)	2.783(13)	–	3.747(10)	1.790(17)	2.837(18)	3.819(10)	4.769(6)
Lin. Fit	2.553(18)	2.772(9)	3.660(14)	3.748(6)	–	–	–	–
Quad. Fit	2.525(38)	2.785(9)	3.687(24)	3.749(6)	–	–	–	–
HBχPT Fit.	2.487(31)	–	–	–	–	–	–	–
Exp.	2.455	2.695	3.519	–	1.672	2.766(2)	–	–
[2]	2.467(50)	2.673(17)	3.603(31)	3.704(21)	1.772(7) [6]	2.738(17)	3.779(23)	4.789(27)
[3]	2.460(46)	2.643(75)	3.568(34)	3.658(77)	1.672(12)	2.728(61)	3.735(74)	4.734(32)
[4]	2.481(46)	2.681(48)	3.595(66)	3.679(62)	–	2.764(49)	3.765(65)	4.761(79)
[5]	2.474(66)	2.679(57)	3.610(45)	3.738(40)	–	2.755(61)	3.822(42)	4.796(26)

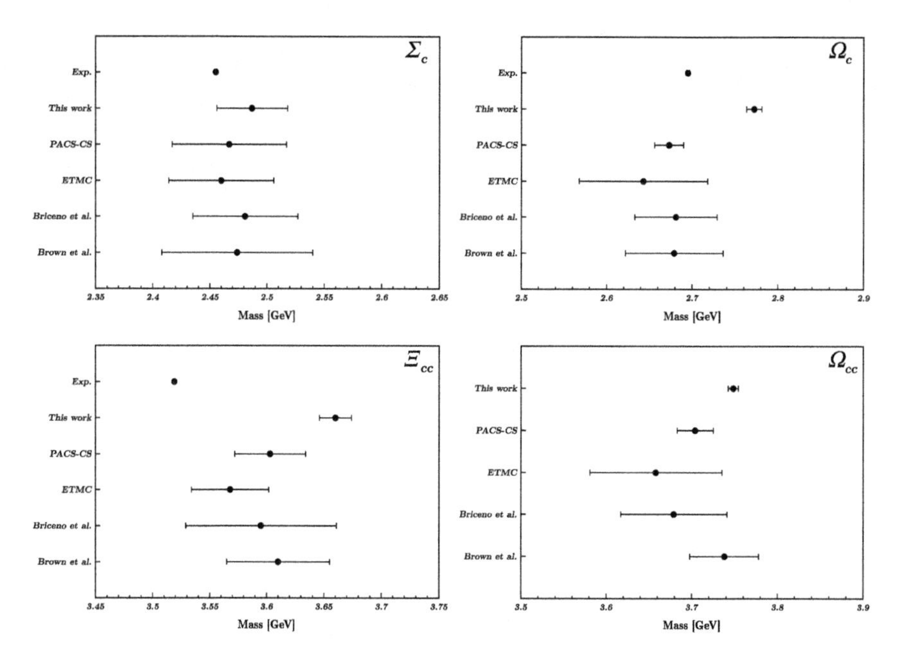

Fig. 5.3 A comparison of the spin-1/2 baryon masses given in Table 5.1

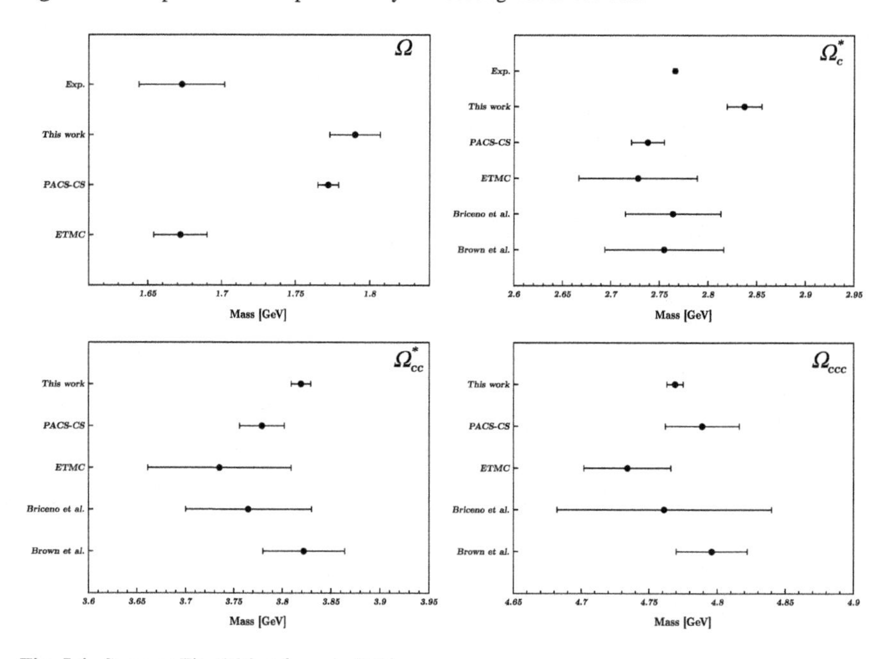

Fig. 5.4 Same as Fig. 5.3 but for spin-3/2 baryons

baryon masses become input parameters via kinematical terms when we evaluate the Eqs. (4.27), (4.28), (4.45), (4.46) and (4.47). We extract the masses of the baryons from the relevant two-point correlation functions given in Eqs. (4.17) and (4.35) as described in Sect. 4.1.1.2 on each ensemble. Correlators are calculated with a Gaussian smeared source and a point sink for better ground state coupling. In case of the spin-3/2 baryons, we extract the masses by a simultaneous fit to all spatial Lorentz components. We show the effective mass plots of each baryon along with the fit regions in Fig. 5.1.

Masses of the spin-3/2 baryons are calculated on the N5 ensemble, which has almost physical light-quark masses, so we consider those values as final. For spin-1/2 baryons, on the other hand, we extrapolate the values to the physical-mass point by using the linear or quadratic functions in m_π^2,

$$f_{\text{lin}} = c_1 + a_1\, m_\pi^2, \tag{5.1}$$

$$f_{\text{quad}} = c_2 + a_2\, m_\pi^2 + b_1\, (m_\pi^2)^2, \tag{5.2}$$

where $a_{1,2}$, b_1, and $c_{1,2}$ are free fit parameters. Extrapolations are shown in Fig. 5.2 and all the extracted masses along with the extrapolated results are given in Table 5.1. Figures 5.3 and 5.4 illustrate a comparison of our values to that of other lattice QCD determinations and experimental masses where available.

Accounting the systematic errors: It is interesting to compare our results for the baryon masses with those obtained by PACS-CS from the same lattices. It must, however, be noted that PACS-CS uses a relativistic heavy-quark action for the c-quark to keep the $\mathcal{O}(m_Q a)$ errors under control and extracts the masses at the physical point without any chiral extrapolation. Such differences between two analyses need to be taken into account as a source of systematic error. Yet, a mass determination, of course, requires a more systematic chiral fit than linear or quadratic forms as we perform here. Nevertheless, such a comparison is useful to see the effect of the discretisation errors in our analysis. The extrapolations to the physical-mass point in linear and quadratic forms are consistent with each other within their error bars. For all baryons, we either see an agreement within error bars or only a few percent discrepancy in baryon masses between PACS-CS and our results. This suggests that the discretisation errors are relatively small.

We perform an additional extrapolation via a Heavy Baryon Chiral Perturbation Theory (HBχPT) inspired form [7] for the Σ_c baryon. With the assumption of heavy quark symmetry, such that $\Sigma_c^* \equiv \Sigma_c$, the extrapolation function is given as

$$M_\Sigma = M_0 + \Delta_{\Lambda\Sigma} - \frac{\sigma_\Sigma}{4\pi f_\pi} m_\pi^2 - \frac{2}{3}\frac{g_{\pi\Lambda\Sigma}^2}{(4\pi f_\pi)^2}\mathcal{F}(m_\pi, -\Delta_{\Lambda\Sigma}, \mu) + \frac{4}{3}\frac{g_{\pi\Sigma\Sigma}^2}{(4\pi f_\pi)^2}\mathcal{F}(m_\pi, 0, \mu), \tag{5.3}$$

where the M_0, σ_Σ, $g_{\pi\Lambda\Sigma}$ and $g_{\pi\Sigma\Sigma}$ are free fit parameters corresponding to bare mass, a low energy constant and $\pi\Lambda_c\Sigma_c$ and $\pi\Sigma_c\Sigma_c$ coupling constants, respectively. Σ_c-Λ_c mass splitting is defined as $\Delta_{\Lambda\Sigma}$, $f_\pi = 132$ MeV and the chiral $\mathcal{F}(a, b, c)$ is [8],

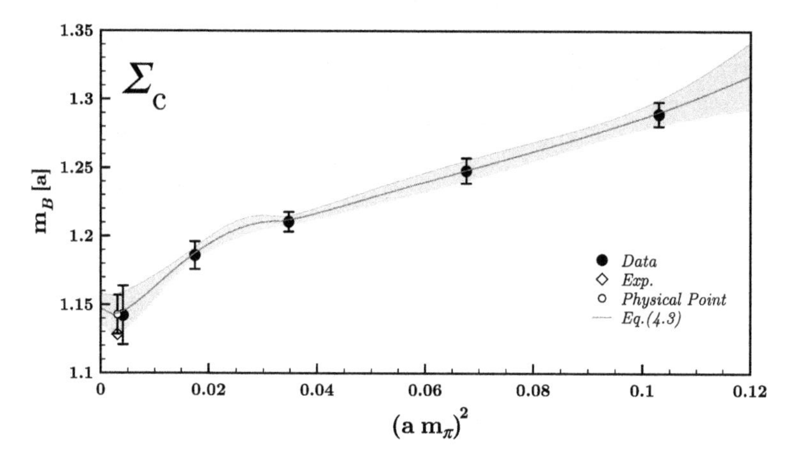

Fig. 5.5 Extrapolation to the physical-mass point using the function given in Eq. (5.3) for the mass of Σ_c baryon. Experimental result is converted to lattice units

$$\mathcal{F}(a,b,c) = (b^2 - a^2)^{3/2} \ln\left(\frac{b + \sqrt{b^2 - a^2}}{b - \sqrt{b^2 - a^2}}\right) - \frac{3}{2} b\, a^2 \ln\left(\frac{a^2}{c^2}\right) - b^3 \ln\left(\frac{4b^2}{a^2}\right), \tag{5.4}$$

$$\mathcal{F}(a,0,c) = \pi a^3, \tag{5.5}$$

$$\mathcal{F}(a,-b,c) = \begin{cases} -\mathcal{F}(a,b,c) + 2i\pi(b^2 - a^2)^{3/2}, & a < |b| \\ -\mathcal{F}(a,b,c) + 2\pi(b^2 - a^2)^{3/2}, & a > |b| \end{cases}. \tag{5.6}$$

Extrapolation is illustrated in Fig. 5.5. HBχPT form yields $m_{\Sigma_c} = 2.487(31)$ GeV, in good agreement with other lattice determinations and its experimental value. Comparing the mean value to the experimental one, a 32 MeV difference translates into less than 2% systematic error on Σ_c baryon masses extracted on ensembles N1–N5.

Effects of the possible mistuning of the mass of the charm quark or the use of Clover action may be investigated by mildly changing the κ_c from its determined value and as long as we confirm that the deviations do not affect the observables that we extract, we can deem the systematic errors due to charm-quark as under control.

We have reasonable expectations that the systematic effects to be much smaller in case of the form factors which are less sensitive to the charm-quark mass [9]. In order to confirm this statement we have explicitly checked the sensitivity of Ξ_{cc} form factors on charm-quark hopping parameter for the N1 ensemble by changing the κ_c so that Ξ_{cc} mass deviates approximately by 100 MeV. Electric charge radii are affected by less than 2%, which is smaller compared to the statistical precision of our electromagnetic observable results that we will present in Sect. 5.3.1.

Another cross-check for the aptness of the use of Clover action for charm quarks is to compare the mass of the Ω_{ccc} baryon to the masses determined by other collaborations which employ different actions to account for the $\mathcal{O}(m_Q a)$ errors and either perform continuum and mass extrapolations to the physical point or makes calculations on the physical-mass point. Since Ω_{ccc} consists of three charm valence quarks, it is a good laboratory to observe the effects of the discretisation or tuning errors. It is apparent in Fig. 5.4 that our Ω_{ccc} mass has very good agreement with other determinations indicating that the systematic errors associated with the c quark is at a minimum.

Note that the differences may also arise from our choice of the strange quark hopping parameter. In order to avoid the partial-quenching effects we have chosen κ_s to be the same as that of the sea quark. On the other hand, our Ω mass is in good agreement with the mass reported by the PACS-CS Collaboration [6]. A retuning of κ_s so as to obtain the physical K mass would be desirable for precision calculations. However we expect such a retuning to have a minimal effect on the conclusions of this work.

It is safe to assume that any systematic uncertainty arising due to our choice of the Clover action or κ_s or κ_c is well under control and negligible with respect to the statistical precision of this work.

5.2 Evaluation of Form Factor Data

5.2.1 Form Factor and Excited-State Analysis

We extract the form factor values corresponding to each momentum insertion by performing fits to Eqs. (4.16) and (4.44) as described in Sect. 4.1.3. We consider three different fit procedures which we discuss in detail below.

5.2.1.1 Plateau Method

In the large Euclidean time limit, ratios are free from excited states however in the actual simulations we should be aware of the possibility that the time separation between the source and the sink particles may not be large enough to allow for the excited states to diminish. Effects of the excited states can be reduced to a minimum by choosing a large enough source-sink separation when computing the correlation functions. When the separation is large enough, ratios in Eqs. (4.16) and (4.44) exhibit regions where their values remain constant with respect to the current insertion time. These regions are identified as *plateau* regions and the value of the form factor is extracted by fitting the data to a constant form,

$$R(t_2, t_1) = R_G + \mathcal{O}(e^{-\Delta t_1}) + \mathcal{O}(e^{-\Delta'(t_2 - t_1)}), \tag{5.7}$$

where the R_G is the ground state value of the ratio and the actual fit function is followed by the terms denoting the first excited state contributions which are suppressed proportional to the current insertion time, t_1, and its difference from the sink time, $(t_2 - t_1)$. Δ and Δ' are the energy gap between the ground and first excited state of the source and sink baryons respectively. It is evident that an ill-identified fit window, $[t_i, t_f]$, or *plateau* might be contaminated by excited states.

5.2.1.2 Phenomenological Fit Form

We can improve the plateau method by adding the higher-order terms to the fit form which explicitly account for the excited states. By including the first excited state, we write down the so-called phenomenological fit form,

$$R(t_2, t_1) = R_G + b_1 e^{-\Delta_1 t_1} + b_2 e^{-\Delta_2 (t_2 - t_1)}, \tag{5.8}$$

where the first term is the form factor value we wish to extract and the coefficients b_1, b_2 and the mass gaps Δ_1, Δ_2 are regarded as free parameters. Regression analysis is performed on the whole set of data, i.e. $[t_i, t_f] = [t_0, t_2]$ where t_0 is the time slice of the source point. In case of the nucleon form factors using the sequential-source inversion method, this approach has proved to be useful in a more systematic analysis accounting for the excited-state contaminations (e.g. Ref. [10] for a rigorous test). More terms of the same order can be added to the above equation to account for further fluctuations, however, keep in mind that there might not be enough degrees of freedom. It is better to keep the number of free fit parameters low if the data fluctuates mildly.

We utilize the phenomenological form as a cross check rather than the actual fit procedure since regression analysis has a tendency to become unstable with increased number of free parameters. As long as the plateau fit results agree with that of the phenomenological form fits, we deem the data as reliable, less prone to excited state contamination and thus, trust the identified plateaux and adopt its values for form factors.

5.2.1.3 Summed Operator Insertions (SOI)

Another strategy that is used to decrease the excited-state contaminations is to vary the source-sink separation and extract the ground state matrix elements by using the *summed operator insertions* method [11]. In the SOI method, one sums the ratio in t_1 up to t_2 so that it assumes the form,

$$\sum_{t_1=0}^{t_2} R(t_2, t_1; \mathbf{p}', \mathbf{p}; \Gamma) = R_G . t_2 + c(\Delta, \Delta') + \mathcal{O}\left(t_2 e^{-\Delta t_2}\right) + \mathcal{O}(t_2 e^{-\Delta' t_2}), \tag{5.9}$$

where the $c(\Delta, \Delta')$ is a constant. First excited-state contributions are now suppressed by t_2, which is larger than t_1 or $(t_2 - t_1)$. It is possible to calculate the ratio with different source-sink separations and extract the ground state value R_G from the slope of a linear function in t_2. This method has the advantage of computing matrix elements with reduced excited-state contaminations, however it is computationally more demanding as calculations with multiple source-sink separations are needed.

Rather than employing this approach as our primary analysis we make use of this method as a cross-check in contrast to the results that we extract by the plateau and phenomenological methods.

5.2.2 Charge Radii and Magnetic Moments

5.2.2.1 Charge Radii

We have argued in Sect. 2.4 that by working in the Breit frame we can expand the Fourier transformation of the EM form factors as,

$$G_{E,M}(Q^2) = \int d^3\mathbf{x} e^{i\mathbf{x}\mathbf{q}} \rho(\mathbf{x}) \simeq G_{E,M}(0)\left(1 - \frac{1}{6} Q^2 \langle r^2_{E,M}\rangle + \dots\right), \tag{5.10}$$

where $\rho(\mathbf{x})$ is a spatial electric charge or magnetisation density distribution. Evaluating the slope of the form factor at $Q^2 = 0$ we can extract the charge radii of the baryons via,

$$\langle r^2_{E,M}\rangle = -\frac{6}{G_{\mathrm{E,M}}(0)} \frac{d}{dQ^2} G_{E,M}(Q^2)\bigg|_{Q^2=0}. \tag{5.11}$$

On the lattice we end up with discrete data points so it is evident that we need to estimate the form factors by a functional form. We use the following ansatz known as a *dipole form* to describe the Q^2 dependence of the baryon form factors:

$$G_{E,M}(Q^2) = \frac{G_{E,M}(0)}{(1 + Q^2/\Lambda^2_{E,M})^2}. \tag{5.12}$$

It is well known that the dipole approximation gives a good description of experimental electric form factor data of the proton. To evaluate the charge radii with the above formula, we insert the dipole form into Eq. (5.11), which yields,

$$\langle r^2_{E,M}\rangle = \frac{12}{\Lambda^2_{E,M}}. \tag{5.13}$$

Then the charge radii can be directly calculated using the values of dipole masses, $\Lambda_{E,M}$, as obtained from our simulations.

5.2.2.2 Magnetic-Dipole Moment

In order to estimate the magnetic moment we need the $Q^2 = 0$ value of the magnetic form factor G_M. However, due to its definition—Eq. (4.28)—we can not compute the $G_M(0)$ value directly since it vanishes at zero momentum transfer. Hence, we obtain its value by extrapolating the lattice data to $Q^2 = 0$ via the dipole form given in Eq. (5.12). In spin$-3/2 \rightarrow$ spin$-3/2$ transition, on the other hand, we estimate the $Q^2 = 0$ value differently since we have calculated only the $G_{M1}(Q^2 = 2\pi/L)$ value of the magnetic form factor. In this case, we assume that the momentum dependence of the $E0$ and $M1$ form factors are similar in the low-Q^2 region. For instance, the scaling of G_{M1} is given by,

$$G_{M1}(0) = G_{M1}(Q^2)\frac{G_{E0}(0)}{G_{E0}(Q^2)}, \tag{5.14}$$

where we consider the scaling of quark sectors separately since each sector has a different scaling property. $G_{M1}(0)$ is then constructed via Eq. (5.18). This procedure has been utilised in Refs. [12–15] to study the magnetic form factors of octet and decuplet baryons.

Once we determine the $G_M(0)$, we evaluate the magnetic moments in nuclear magnetons using the relation,

$$\mu_{\mathcal{B}} = G_M(0)\left(\frac{e}{2m_{\mathcal{B}}}\right) = G_M(0)\left(\frac{m_N}{m_{\mathcal{B}}}\right)\mu_N, \tag{5.15}$$

where m_N is the physical nucleon mass and $m_{\mathcal{B}}$ is the baryon mass as obtained on the lattice.

5.2.2.3 Electric-Quadrupole Moment

Higher orders in the multipole expansion of the form factors can be evaluated given that they are allowed by the angular momentum selection rule. Let us consider the electromagnetic transition, $\langle J'|J_\gamma|J\rangle$, where $J^{(\prime)} = L^{(\prime)} + S^{(\prime)}$ is the total angular momentum of the initial (final) state and J_γ is the total angular momentum of the photon. The selection rule we have states,

$$|J - J'| \leq J_\gamma \leq J + J'. \tag{5.16}$$

Considering the ground state spin$-1/2 \rightarrow$ spin$-1/2$ electromagnetic transition we have at most two form factors, which we have identified as Dirac and Pauli or electric and magnetic Sachs form factors in Sect. 4.1.2.1. Higher order multipole form factors vanish when $L = 0$.

In case of the spin$-3/2 \rightarrow$ spin$-3/2$ transition, selection rule allows us to have four multipole form factors contributing to the interaction which we have given in

Sect. 4.1.2.2. The electric-quadrupole form factors provide information about the shape of the electric charge distribution of the baryon and a non-zero value is also an indication of tensor force interactions. In the Breit frame, the quadrupole form factor and the electric charge distribution are related as [12],

$$\mathcal{G}_{E2}(0) = M_B^2 \int d^3r \bar{\psi}(r)(3z^2 - r^2)\psi(r), \tag{5.17}$$

where $3z^2 - r$ is the quadrupole moment operator. A non-zero value of the quadrupole moment indicates a deviation from a spherically symmetric charge distribution and the shape is determined by the sign of the moment where a positive (negative) value is assigned to a prolate (oblate) shape for a positively charged baryon. We use Eq. (4.46) to isolate the electric-quadrupole form factor and estimate the quadruple moment.

5.2.3 *Quark Sectors*

In our simulations we evaluate each quark sector separately and normalise to unit charge contributions. Baryon properties, on the other hand, are determined by the collective contribution of individual quarks so we estimate the baryon observables by combining the appropriately weighted quark sector contributions as,

$$\langle \mathcal{O} \rangle = N_{u,d} e_{u,d} \langle \mathcal{O}_{u,d} \rangle + N_s e_s \langle \mathcal{O}_s \rangle + N_c e_c \langle \mathcal{O}_c \rangle, \tag{5.18}$$

where $\langle \mathcal{O} \rangle$ is the observable, N_q is the number of quarks inside the baryon having flavor q and e_q is the electric charge of the quark. Difference in the analysis might introduce systematical errors at this stage, but the considerations we make in Sect. 5.5.3 shows that the case is the opposite.

5.3 Spin-1/2 Baryons

We have performed simulations and extracted the electromagnetic form factors of the spin-1/2 Σ_c, Ω_c, Ξ_{cc} and Ω_{cc} baryons. Theoretical formalism is given in Sect. 4.1.2.1. Electromagnetic form factors of the spin$-1/2 \rightarrow$ spin$-1/2$ transition are related to the electric charge and magnetisation radii and the magnetic moments of the spin-1/2 baryons. In the following sections we will give the numerical values of the form factors, charge radii and magnetic moments and discuss the physical implications.

5.3.1 Electric Properties

5.3.1.1 Plateau Analysis

We study the electric Sachs form factor, $G_E(Q^2)$, given in Eq. (4.27) and extract the electric charge radii of the baryons in question. We have evaluated the form factor values corresponding to each momentum insertion that we consider by a plateau method analysis as described in Sect. 5.2. Figure 5.6 shows the ratio in Eq. (4.16) as a function of the current insertion time, t_1, for the electric form factors of Σ_c^{++} and Ω_{cc}^{+} as normalized with their electric charges. We present the data solely for the ensemble N1 and for the first nine four-momentum insertions. In order to illustrate how plateau regions change as we approach to the physical point, we show the electric form factors of Ξ_{cc}^{++} in Fig. 5.7 for the ensembles N1–N4.

In determining a plateau region we vary the fit window $[t_i, t_f]$ and compare the fits by their χ^2 values. For instance we search for plateau regions of minimum three time slices between the source and the sink, and choose the one that has the best χ^2. We rely on this method since the signal exhibits a different ground state approach because we use an asymmetric (Gaussian smeared) source-(wall smeared) sink pair. Wall smearing has a weak coupling to the ground state compared to the Gaussian smearing, therefore making the data closer to the sink time slice more prone to excited-state contamination. We prefer the regions closer to the smeared source as we expect them to couple to the ground state with higher strength as compared to the wall sink.

In our simulations, the source-sink time separation is fixed to ~1.09 fm ($t_2 = 12a$). The signal-to-noise ratio deteriorates as we increase the seperation. Increasing the number of measurements as a means of increasing the statistics would improve the signal, however, it would also increase the cost of the computations and enough

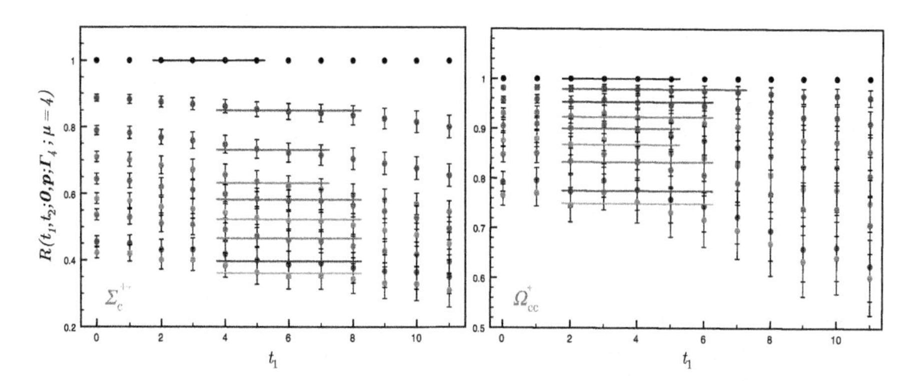

Fig. 5.6 The ratio in Eq. (4.16) as a function of the current insertion time, t_1, for the electric form factors of Σ_c^{++} and Ω_{cc}^{+} as normalized with their electric charges. We show the data of the ensemble N1 only. Horizontal lines denote the plateau regions as determined by using a χ^2 criterion (see text). Figure taken from Ref. [16]

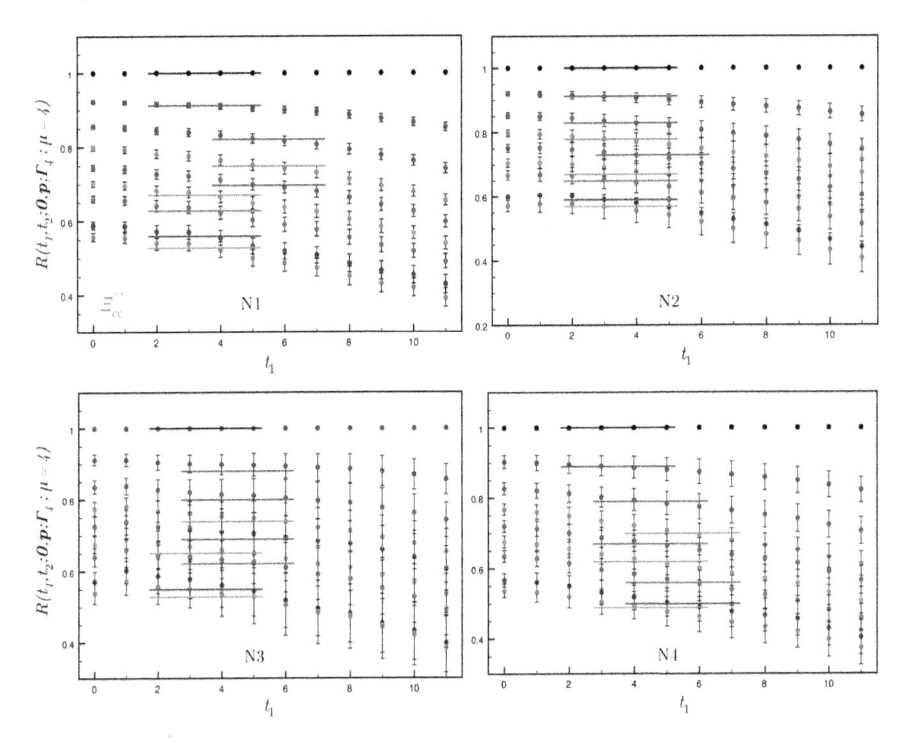

Fig. 5.7 Same as Fig. 5.6 but for the electric form factors of Ξ_{cc}^{++} obtained on ensembles N1–N4. Figure taken from Ref. [16]

resources (e.g. gauge configurations) might not always be available. Hence, effectively, we have an upper limit for the value of t_2. Therefore, we must choose the smallest possible separation value ensuring that the excited-state contaminations are avoided. There are works in the literature that finds a separation of ~1 fm sufficient for the nucleon axial and electromagnetic form factors [17, 18]. In case of the strange sector a similar conclusion has also been made for the Ω^- electromagnetic form factors [19]. However, there has not been any work on the charmed baryon electromagnetic form factors by the lattice community so we check that whether a separation of $t_2 = 12a$ is sufficient for the charmed baryons.

We compute the same observables using a separation of $t_2 = 14a$ and compare our results with those obtained using a separation of $t_2 = 12a$. In Fig. 5.8 we show the ratio in Eq. (4.16) to illustrate a test case for the electric form factor of Ξ_{cc} with $t_2 = 12a$ and $t_2 = 14a$. Data points clearly indicate the the plateau values obtained from the two time separations are consistent with each other, implying that the shorter source-sink time separation is sufficient. The error bars for $t_2 = 14a$, however, are at least twice as large compared to $t_2 = 12a$. It would require us to at least double the number of measurements to reach a similar precision of the $t_2 = 12a$ case. This is unfortunately not possible since we already exploit the translational symmetry to increase the number of source-sink pairs, hence the number of measurements, and

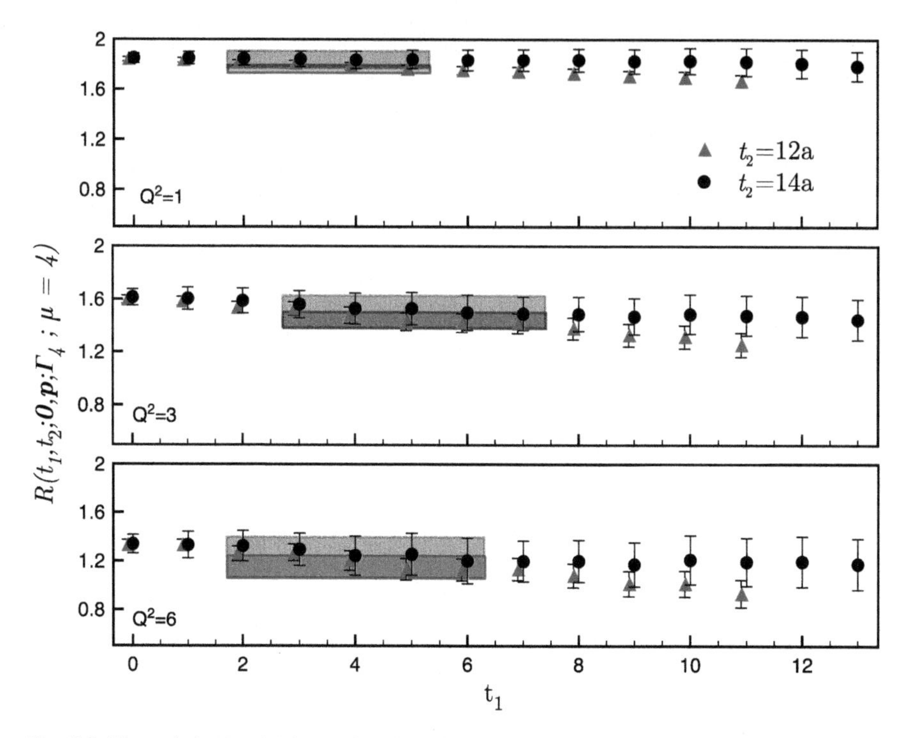

Fig. 5.8 The ratio in Eq. (4.16) as a function of the current insertion time, t_1, for the electric form factor of Ξ_{cc} with $t_2 = 12a$ and $t_2 = 14a$. We show measurements over 30 configurations for three illustrative momentum-transfer values. The data for $t_2 = 12a$ are slightly shifted to left for clear viewing. Figure taken from Ref. [16]

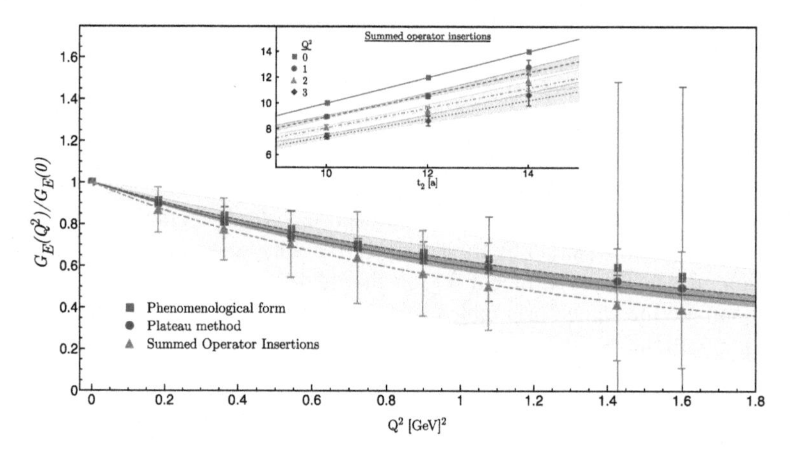

Fig. 5.9 A comparison of the electric form factor of Ξ_{cc} as obtained on ensemble N1 using a simple plateau fit, the phenomenological fit form in Eq. (5.8) and the summation method. The small panel depicts the summed operator insertions for three time separations and for the first four momentum insertions with their linear fits. Figure taken from Ref. [16]

Table 5.2 Parameter values of $R(t_2, t_1)$ in case of the electric form factors of Ξ_{cc} for all momentum transfers on ensemble N1

Q^2	$G_E(Q^2)$		Δ	b_1	b_2
$[\frac{2\pi}{aN_s}]$	Plat.	Pheno.	[a]		
1	0.913(7)	0.910(18)	0.220(100)	0.026(18)	−0.087(18)
2	0.822(13)	0.839(42)	0.186(90)	0.045(34)	−0.143(35)
3	0.750(15)	0.775(85)	0.169(110)	0.064(56)	−0.174(71)
4	0.698(16)	0.703(27)	0.285(108)	0.082(33)	−0.154(22)
5	0.671(15)	0.664(53)	0.212(122)	0.079(47)	−0.171(41)
6	0.629(16)	0.634(202)	0.174(160)	0.078(113)	−0.197(169)
8	0.560(19)	0.594(889)	0.145(251)	0.062(418)	−0.210(672)
9	0.529(20)	0.552(908)	0.160(356)	0.066(466)	−0.205(700)

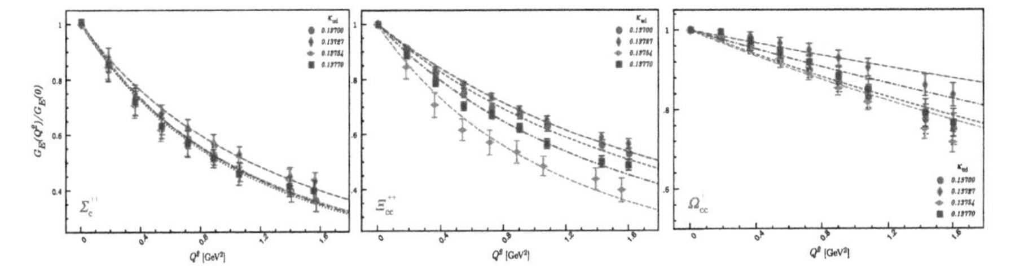

Fig. 5.10 The electric form factors of Σ_c^{++},Ξ_{cc}^{++} and Ω_{cc}^{+} as normalised with their electric charges as functions of Q^2, for all the quark masses we consider. The dots mark the lattice data and the curves show the best fit to the dipole form in Eq. (5.12). Figure taken from Ref. [16]

also since the number of gauge configurations available is limited. Other baryons we study exhibit a similar behaviour, therefore we use the shorter separation i.e. $t_2 = 12a$, in all of our analyses.

Still, we have to ensure that the excited-state contamination is at a minimum in the plateau regions that we have chosen. We perform fits using the phenomenological fit form for the Ξ_{cc} baryon as a test case. In Fig. 5.9 we present a comparison of the electric form factor of Ξ_{cc} as obtained by a plateau fit and the phenomenological fit form in Eq. (5.8), for three illustrative momentum transfers and for the ensemble N1. The two fit forms give completely consistent results although the error bars being twice as large for the phenomenological fit form. We compile the parameter values of $R(t_2, t_1)$ for the electric form factors of Ξ_{cc} in Table 5.2 for all momentum transfers. The statistical error in the Δ values is quite large as expected since we regard it as a free parameter. Note that even though the definition of Δ corresponds to the energy gap between the ground state and the first excited state, we do not intend to interpret Δ as the physical energy gap at this stage since this way of determining such a parameter is not viable compared to the advanced methods utilised for spectroscopy.

Agreement between the plateau and phenomenological fit method is satisfactory but to further check that we avoid excited-state contaminations, we can take one more step and compute the ratio in Eq. (4.16) for $t_2 = 10a$ source-sink separation also. This would give us a data set containing ratio values for three source-sink separations, namely for $t_2 = 10a$, $t_2 = 12a$ and $t_2 = 14a$ (on ensemble N1 with 30 configurations). We use the summed operator insertions method on this data set to extract the form factors and compare the results to that of other methods. Figure 5.9 also depicts a comparison of the summation method with the other methods. In general, statistical errors of the summation method are larger, however, data are consistent within the errors leading us to conclude that the excited state contaminations are negligible.

5.3.1.2 Electric Charge Radii

Once we extract the form factor values, we estimate the electric charge radii of the baryons by Eq. (5.13). First, we fit the form factor data by a dipole form given in Eq. (5.12) to extract the dipole mass variable, Λ_E. Figure 5.10 displays the electric form factors of Σ_c^{++}, Ξ_{cc}^{++} and Ω_{cc}^{+}, normalised by their electric charges, as functions of Q^2. We show the lattice data and the fitted dipole forms obtained on the ensembles N1–N4. The dipole form describes the lattice data for the charmed baryons quite successfully also, with high-quality fits.

We, then, simply use the Λ_E values to evaluate the Eq. (5.13) and estimate the electric charge radii of the Σ_c^{++}, Ξ_{cc}^{+}, Ξ_{cc}^{++} and Ω_{cc}^{+} baryons. Our numerical results are given in Table 5.3. We give the electric charge radii in fm^2, as calculated on ensembles N1–N4. These numerical values are illustrated in Fig. 5.11 with their extrapolations to the physical-mass point for the electric radii. To obtain the values of the observables at the physical point, we perform fits that are constant, linear and quadratic in m_π^2:

Table 5.3 The electric charge radii of Σ_c^{++}, Ξ_{cc}^{+}, Ξ_{cc}^{++} and Ω_{cc}^{+} as obtained on ensembles N1–N4 and extrapolated values to the physical-mass point. Charge radii are given in units of fm^2. χ^2/dof and p-values of the linear and quadratic fits are also given

ID	$\langle r^2_{E,\Sigma_c^{++}} \rangle$	$\langle r^2_{E,\Xi_{cc}^{+}} \rangle$	$\langle r^2_{E,\Xi_{cc}^{++}} \rangle$	$\langle r^2_{E,\Omega_{cc}^{+}} \rangle$
N1	0.206(23)	0.035(6)	0.118(8)	0.038(8)
N2	0.170(19)	0.017(5)	0.107(6)	0.019(6)
N3	0.196(27)	0.018(6)	0.127(8)	0.040(6)
N4	0.195(34)	0.032(8)	0.142(9)	0.029(6)
Lin. Fit	0.192(22)	0.017(6)	0.136(8)	0.032(6)
χ^2/dof	0.916	3.183	2.887	3.701
p-Val	0.40	0.04	0.06	0.025
Quad. Fit	0.234(37)	0.042(9)	0.165(12)	0.043(11)
χ^2/dof	0.222	0.187	0.177	5.988
p-Val	0.64	0.66	0.67	0.014

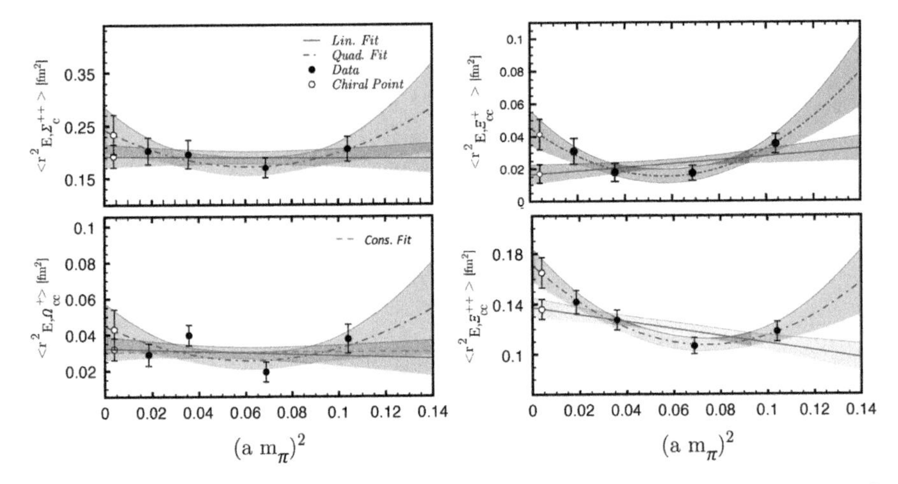

Fig. 5.11 Chiral extrapolations for electric charge radii of Σ_c^{++}, Ξ_{cc}^{+}, Ξ_{cc}^{++} and Ω_{cc}^{+} in $(am_\pi)^2$. We show the fits to constant, linear and quadratic forms. Shaded regions are the maximally allowed error regions, which give the best fit to data. Figure taken from Ref. [16]

$$f_{\text{cons.}} = c_1, \tag{5.19}$$

$$f_{\text{lin}} = a_1\, m_\pi^2 + b_1, \tag{5.20}$$

$$f_{\text{quad}} = a_2\, m_\pi^4 + b_2\, m_\pi^2 + c_2, \tag{5.21}$$

where $a_{1,2}, b_{1,2}, c_{1,2}$ are the free fit parameters. Extrapolations with linear and quadratic forms deviate from each other with their one to two standard deviations in some cases, in particular for Σ_c. We also give a goodness of fit analysis of our extrapolations for different fit forms in Table 5.3. A closer inspection with the χ^2 per degree of freedom and the p-values taken into account reveals that the quadratic form is favoured in case of the charge radii.

Shifting our attention to the charmed-strange baryon Ω_{cc}^{+}, we expect the pion-mass dependence to be solely due to sea-quark effects since Ω_{cc}^{+} has no light valence quarks. By a closer look to the lowest left panel of Fig. 5.11, we see that the dependence of charge radii of the Ω_{cc} fluctuate as we approach the chiral limit, in contrary to the naive expectation. This fluctuation might just be a statistical fluctuation or may be due to uncontrolled systematic errors as well. Since a functional form incorporating the sea-quark dependence is not known we fit these data to a constant or a linear form. Unfortunately, the fluctuating data results in a poor fit to a linear or quadratic form for the charge radii of the Ω_{cc}^{+} baryon. Note that the data in other cases can be nicely fit to linear or quadratic forms.

We also account for the consistency between the properties of the baryons as extrapolated to the quark-mass point $m_\pi^2 = m_{\eta_{ss}}^2$ in order to assess the best fit function to data. We do not compute the $m_{\eta_{ss}}$ at the SU(3) symmetric point, however, we use the PACS-CS determined value of $m_{\eta_{ss}} = 0.39947$ [6], to make an estimation. we expect the charge radii of $\Xi_{cc}^{+}(dcc)$ and $\Omega_{cc}^{+}(scc)$ to coincide at this point since the mass of

Table 5.4 Individual quark sector contributions to the electric charge radii of the charmed baryons. Note that the numbers are given in units of fm^2 and independently from the electric charge of the individual quarks that compose the baryons

ID	$\Sigma_c^{0,++}$		$\Xi_{cc}^{+,++}$		Ω_c^0		Ω_{cc}^+	
	$\langle r_E^2 \rangle_q$	$\langle r_E^2 \rangle_Q$	$\langle r_E^2 \rangle_q$	$\langle r_E^2 \rangle_Q$	$\langle r_E^2 \rangle_q$	$\langle r_E^2 \rangle_Q$	$\langle r_E^2 \rangle_q$	$\langle r_E^2 \rangle_Q$
N1	0.289(49)	0.091(22)	0.264(29)	0.071(5)	0.253(35)	0.074(20)	0.249(29)	0.071(7)
N2	0.273(41)	0.054(12)	0.282(25)	0.056(5)	0.199(34)	0.049(15)	0.198(18)	0.051(6)
N3	0.353(65)	0.042(18)	0.379(38)	0.063(5)	0.320(28)	0.076(13)	0.276(22)	0.082(6)
N4	0.338(60)	0.057(25)	0.358(38)	0.080(7)	0.313(36)	0.061(10)	0.316(48)	0.074(9)
Lin. Fit	0.347(49)	0.032(18)	0.386(33)	0.068(5)	0.330(32)	0.064(10)	0.287(31)	0.078(7)
Quad. Fit	0.390(86)	0.066(32)	0.410(46)	0.095(9)	0.398(52)	0.069(22)	0.422(51)	0.104(13)

the light-quark would correspond to that of strange quarks. The properties of the Σ_c^{++} baryon as extrapolated to this region on the other hand, can be compared with those of an unphysical baryon similar to Ω_c^{++} but the s quarks are assigned with electric charge 2/3—a state that can be easily created on our setup with trivial replacements. This approach is rather advisory than being deterministic since it assumes that the chiral limit and heavy quark limit behaviours to be similar, which may as well be different.

Considering the quadratic fit values we can compare the electric charge radii of the baryons that have the same electric charge. The charge radii of Ω_{cc}^{+} and Ξ_{cc}^{+} are about the same size, which is much smaller as compared to that of the proton (the experimental value is $\langle r_{E,p}^2 \rangle$ =0.770 fm^2 [20]). The only internal difference between Ω_{cc}^{+} and Ξ_{cc}^{+} is that we change the light quark with the strange quark or vice versa, yet the s quark in Ω_{cc}^{+} seems to have no extra effect on charge radius with respect to the light quark in Ξ_{cc}^{+}. Of all the four charged baryons (Σ_c^{++}, Ξ_{cc}^{+}, Ξ_{cc}^{++} and Ω_{cc}^{+}) we have studied, Σ_c^{++} appears to have the largest charge radius.

Since we compute the observables for each quark sector individually we can examine their contributions to the electromagnetic form factors to gain a deeper insight to the quark dynamics. Table 5.4 displays the radii of light- (u/d, s) and c-quark distributions within the baryons. We clearly see that the light quark distributions are systematically larger than those of the c quark. Smaller values of the c quark suggests that it acts as a heavy core to shift the center of mass towards itself thus reducing the size of the baryon. Comparing the c-quark distributions between the singly and the doubly charmed baryons we see that the difference is small. Similarly, the u/d- and s-quark distributions are roughly the same. In this case, much of the difference arises due to the electric charges and the representation of the valence quarks in the baryon. For instance we see that Ω_{cc}^{+} and Ξ_{cc}^{+} have almost the same sizes whereas the charge radius of Σ_c^{++} is slightly larger than that of Ξ_{cc}^{++} since the doubly represented u quark has larger contribution than the c quark.

Analysing the change in the light-quark mass exposes an interesting effect: As the u/d quark in Σ_c and Ξ_{cc} baryons becomes lighter the radius of the light quark increases. This can be understood by the shift in the center of mass towards the heavy c quark leading the light quark to have a larger distribution. An unexpected behaviour, however, occurs when the mass of the u/d quark increases: Initially, the charge radii decrease but they increase as we approach to the s-quark mass region, a behaviour which is described nicely by a quadratic function. Unlike the nucleon, it is interesting that the charge radii do not systematically decrease as the pion mass increases. Although this seems to contradict with our findings stating that heavier quarks have smaller charge radii, a modification of the confinement force in hadrons might be responsible for this behaviour such that the two charm quarks assume a compact nature in the Ξ_{cc} and the effect of the extra light quark modifies the string tension between the two-charm component [21].

On the other hand, quark sectors of the Ω_{cc}^{+} baryon shows a somewhat unstable quark-mass dependence making it harder to give a firm statement. Remember that the valence s-quark mass is fixed in our calculations so that the variation is solely due to the effects of the u/d quarks in the sea. Fit analyses with linear and quadratic

Table 5.5 Comparison of the electric charge radii of Ξ_{cc}^{+} and Ξ_{cc}^{++} determined with different κ_c. Charge radii are given in units of fm^2. Last two columns should be compared to each other for reliability since the values are extracted with less measurements

ID	$\langle r^2_{E,\Xi_{cc}^{+}}\rangle$		$\langle r^2_{E,\Xi_{cc}^{++}}\rangle$			
	$\kappa_c =$ 0.1224	$\kappa_c =$ 0.1246	$\kappa_c =$ 0.1224	$\kappa_c =$ 0.1246	$\kappa_c =$ 0.1224	$\kappa_c =$ 0.1232
					$N_{meas} = 30$	
N1	0.052(10)	0.035(6)	0.136(12)	0.118(8)	0.154(28)	0.156(28)
N2	0.027(8)	0.017(5)	0.113(8)	0.107(6)	–	–
N3	0.021(8)	0.018(6)	0.120(9)	0.127(8)	–	–
N4	0.037(10)	0.032(8)	0.144(13)	0.142(9)	–	–
Lin. Fit	0.020(7)	0.017(6)	0.135(11)	0.136(8)	–	–
Quad. Fit	0.049(12)	0.042(9)	0.164(18)	0.165(12)	–	–

functions reveal slight sea-quark mass dependences, suggesting possible sea-quark effects.

Note on systematic errors: In Sect. 5.1 we have mentioned that systematic errors due to the tuning of κ_c affects the electromagnetic observables minimally. Prior to the results we are presenting here, we have studied the Ξ_{cc} baryon in detail with a choice of $\kappa_c = 0.1224$, which we had determined by tuning the mass of the J/ψ on the lattice to its experimental value. In the succeeding work (this work), however, we have improved the tuning procedure and re-determined the hopping parameter of the charm quark to be $\kappa_c = 0.1246$ as discussed in Sect. 4.2.2. We can take advantage of this situation to compare the results of these two works to corroborate our statement. Let us note that we will not present the middle steps of the work with $\kappa_c = 0.1224$ but only quote the charge radii results. Further details of that work is same as the ones outlined in the previous sections and can be found in Ref. [22]. We recompile the Ξ_{cc} results into Table 5.5 for a convenient comparison. Results clearly indicate that a mild change in κ_c has negligible effect on the final electromagnetic observables.

Findings for the electric charge radii

- *General features*:
  - Charge radii of all the charmed baryons that we have studied are much smaller compared to that of their lighter counterparts.
  - Σ_c^{++} has the largest charge radius followed by the Ξ_{cc}^{++} baryon.
  - Charge radii of the Ω_{cc}^{+} (scc) and Ξ_{cc}^{+} (dcc) are about the same size implying that exchanging the light quark with the strange quark, or vice versa, has no extreme effect.

- *Quark sectors*:
  - Charm-quark contributions to the charge radi is systematically smaller than those of the light quark suggesting that the charm quark acts as a heavy core and shifts the center of mass towards itself.

5.3.2 *Magnetic Properties*

Data analysis of this section closely resembles that of the electric sector given in the previous section.

5.3.2.1 Plateau Analysis

In case of the magnetic sector, we study the magnetic Sachs form factor, $G_M(Q^2)$, given in Eq. (4.28) and extract the magnetisation radii and the magnetic moments of

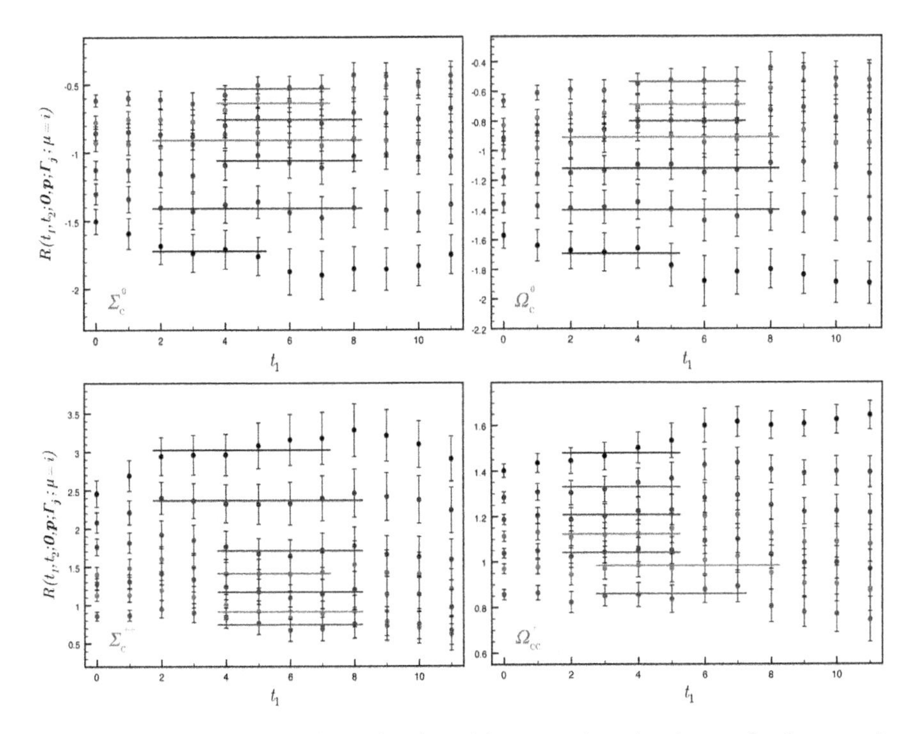

Fig. 5.12 The ratio in Eq. (4.16) as a function of the current insertion time, t_1, for the magnetic form factors of Σ_c^0, Σ_c^{++}, Ω_c^0 and Ω_{cc}^+. We show the ensemble N1 data only. Horizontal lines denote the plateau regions as determined by using a χ^2 criterion, same as the electric sector. Figure taken from Ref. [16]

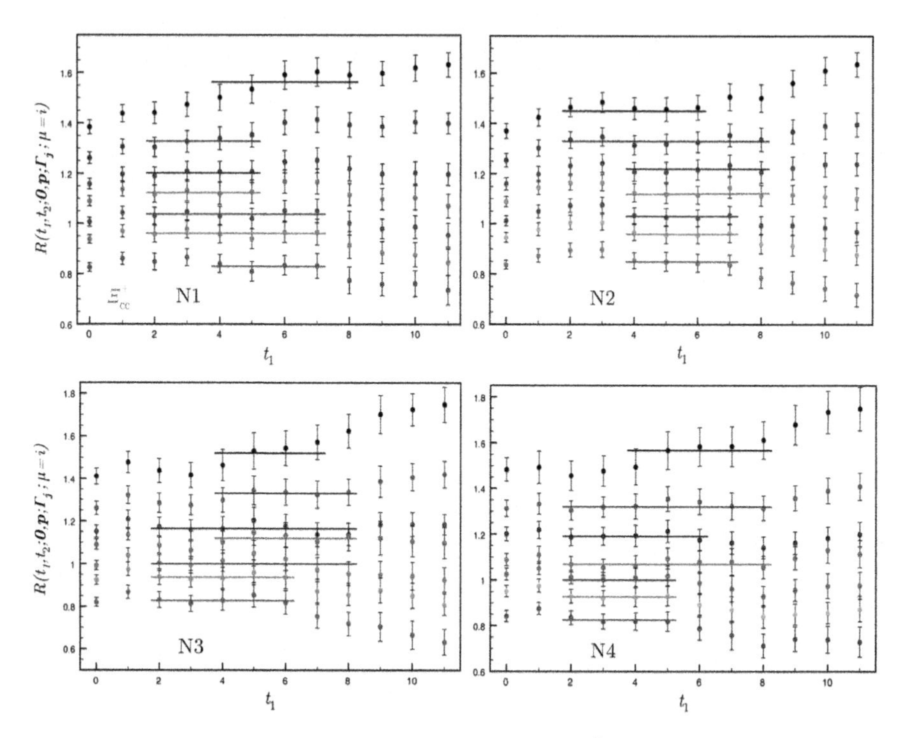

Fig. 5.13 Same as Fig. 5.6 but for the magnetic form factors of Ξ_{cc}^{+} obtained on ensembles N1–N4. Figure taken from Ref. [16]

the baryons that we investigate. Similar to our analysis in the electric form factor sector, we extract the form factor values by the plateau method approach. Figure 5.12 shows the ratio in Eq. (4.16) for the magnetic form factors of the Σ_c^0, Σ_c^{++}, Ω_c^0 and Ω_{cc}^{+} baryons as obtained on ensemble N1. Behaviour of the ratio as the quark mass decreases is presented in Fig. 5.13 where we show the data from ensembles N1–N4.

Remember that the source-sink seperation is fixed to ~ 1.09 fm ($t_2 = 12a$) in our simulations and we did check the electric form factor results by increasing the seperation to ~ 1.26 fm ($t_2 = 14a$). We make a similar test for the magnetic form factors also. In Fig. 5.14 we show the ratio in Eq. (4.16) to illustrate a test case for the magnetic form factor of Ξ_{cc} with $t_2 = 12a$ and $t_2 = 14a$ seperations. Behaviour of the data points on each time seperation is consistent so that we conclude the shorter source-sink time separation is sufficient for the magnetic sector as well. This simple check is satisfactory to confirm that the excited state contaminations are under control in our plateau analysis.

5.3.2.2 Magnetisation Radii and Magnetic Moment

We plot the magnetic form factors in Fig. 5.15 as functions of Q^2 as obtained on ensembles N1–N4. Magnetisation radii are estimated by evaluating the Eq. (5.13) where the dipole mass parameter, Λ_M, is extracted via performing dipole-form—Eq. (5.12)—fits to the form factors. We compile the magnetisation radii into Table 5.6. In Fig. 5.16, we show the charge radii and extrapolations to the physical-mass point. Extrapolating functions are given in Eq. (5.19). A similar pattern to the electric sector can be seen for the magnetisation radii of the charmed baryons also. Quadratic form gives a better fit for all baryon magnetisation radii. Σ_c^{++} has the largest magnetisation radii. Σ_c^{++} and Σ_c^0 seem to have a similar magnetic radii to that of the proton, which is $\langle r^2_{M,p}\rangle = 0.604$ fm^2 [20]. Ω_{cc} has the smallest magnetisation radii.

Magnetic moments of the baryons are estimated via the Eq. (5.15). We obtain the $G_M(0)$ by extrapolating the lattice data to $Q^2 = 0$ via the dipole form in Eq. (5.12). In Table 5.7 we give the $G_M(0)$ values and the magnetic moments evaluated on the ensembles N1–N4. Magnetic form factors and their extrapolations are illustrated

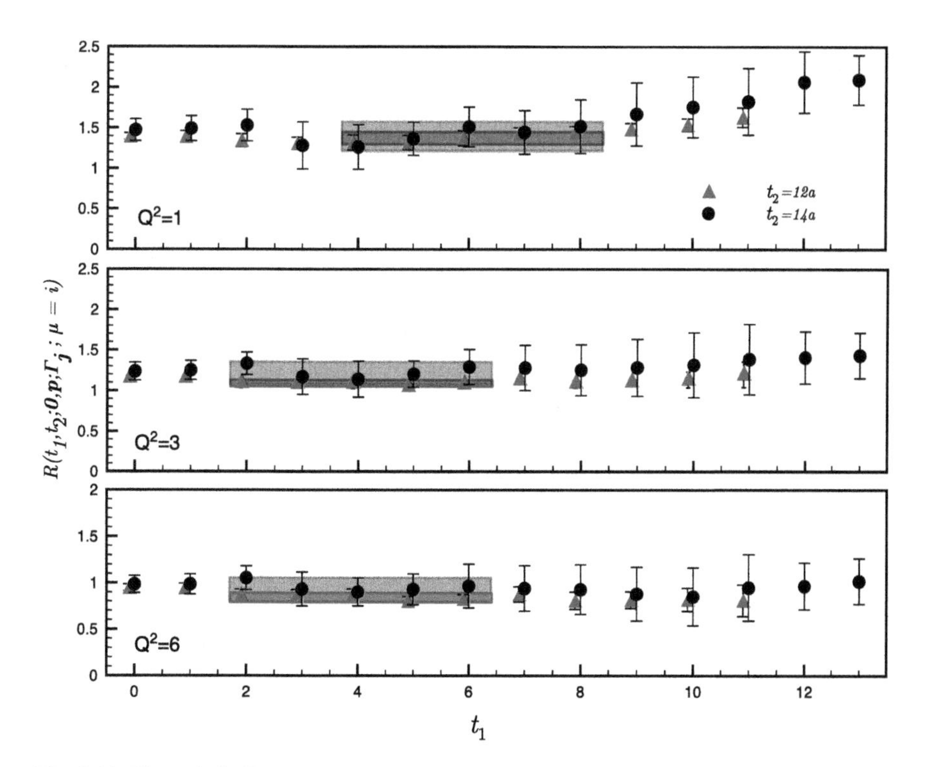

Fig. 5.14 The ratio in Eq. (4.16) as a function of the current insertion time, t_1, for the magnetic form factor of Ξ_{cc} with $t_2 = 12a$ and $t_2 = 14a$. We show statistics over 30 configurations for three illustrative momentum-transfer values. The data for $t_2 = 12a$ are slightly shifted to left for clear viewing. Figure taken from Ref. [16]

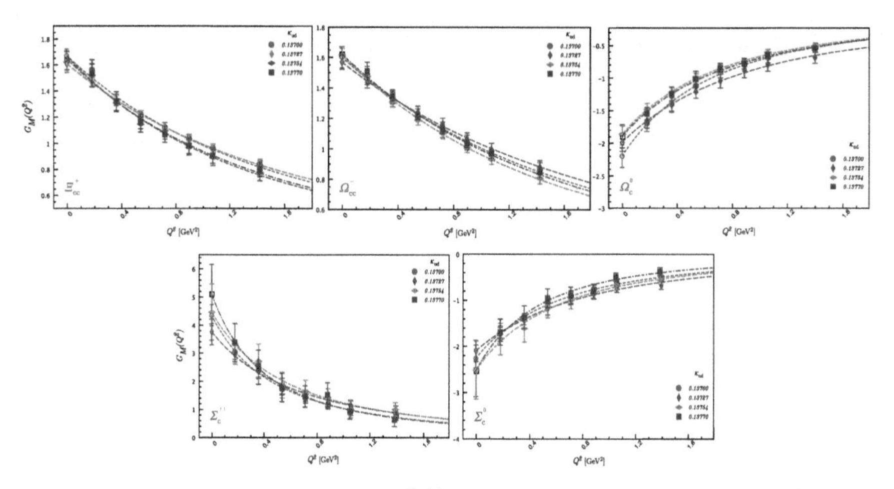

Fig. 5.15 The magnetic form factors of $\Sigma_c^{0,++}$, Ξ_{cc}^+, Ω_c^+ and Ω_{cc}^+ as functions of Q^2, for the ensembles N1–N4. The dots mark the lattice data and the curves show the best fit to the dipole form in Eq. (5.12). Figure taken from Ref. [16]

Table 5.6 The magnetisation radii of $\Sigma_c^{0,++}$, Ξ_{cc}^+, Ω_c^0 and Ω_{cc}^+ as obtained on ensembles N1–N4 and the extrapolated values to the physical-mass point. Charge radii are given in units of fm^2. χ^2/dof and p-values of the linear and quadratic fits are also given

ID	$\langle r^2_{M,\Sigma_c^0}\rangle$	$\langle r^2_{M,\Sigma_c^{++}}\rangle$	$\langle r^2_{M,\Xi_{cc}^+}\rangle$	$\langle r^2_{M,\Omega_c^0}\rangle$	$\langle r^2_{M,\Omega_{cc}^+}\rangle$
N1	0.379(47)	0.492(66)	0.141(9)	0.346(43)	0.122(12)
N2	0.287(44)	0.360(56)	0.127(10)	0.247(39)	0.109(12)
N3	0.391(87)	0.419(77)	0.136(12)	0.313(30)	0.138(11)
N4	0.507(111)	0.574(133)	0.141(13)	0.303(29)	0.130(13)
Lin. Fit	0.377(75)	0.410(81)	0.135(10)	0.297(33)	0.135(11)
χ^2/dof	2.265	1.823	0.613	1.56	1.162
p-Val	0.1	0.27	0.54	0.16	0.31
Quad. Fit	0.650(126)	0.696(53)	0.154(19)	0.354(54)	0.148(21)
χ^2/dof	0.003	0.065	0.091	1.624	1.804
p-Val	0.96	0.69	0.76	0.8	0.18

in Fig. 5.17. Σ_c^{++} has the largest magnetic moment of all and the charmed-strange baryons, Ω_c and Ω_{cc}, have somewhat smaller moments. An inspection of the Ω_c and Ω_{cc} magnetic moments and their dependence on the pion mass, which is due to only sea-quarks, reveals that the moments are almost independent of the sea quark effects. Magnetic moments of all the baryons are smaller in magnitude compared to the experimental magnetic moment of the proton, which is $\mu_p = 2.793\ \mu_N$ [20].

Table 5.8 displays a comparison of our results for the magnetic moments with those from various other models. While the signs of the magnetic moments are

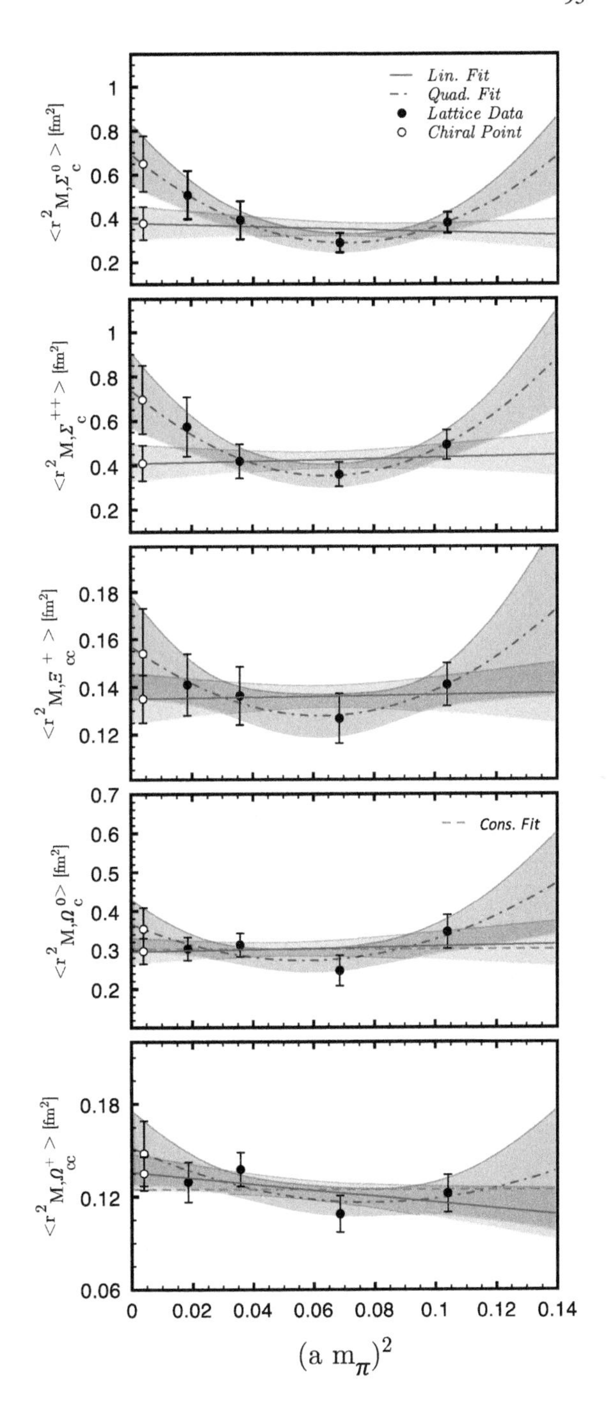

Fig. 5.16 Chiral extrapolations for the magnetisation radii of $\Sigma_c^{0,++}$, Ξ_{cc}^+, Ω_c^0 and Ω_{cc}^+ in $(am_\pi)^2$. We show the fits to constant, linear and quadratic forms. Shaded regions are the maximally allowed error regions, which give the best fit to data. Figure taken from Ref. [16]

Table 5.7 Magnetic form factor values at $Q^2 = 0$ and the magnetic moments of $\Sigma_c^{0,++}$, Ξ_{cc}^+, Ω_c^0 and Ω_{cc}^+ as obtained on ensembles N1–N4. Lower panel holds the extrapolated values to the physical-mass point. Magnetic moments are given in units of nuclear magnetons. χ^2/dof and p-values of the linear and quadratic fits are also given.

ID	Σ_c^0		Σ_c^{++}		Ξ_{cc}^+		Ω_c^0		Ω_{cc}^+	
	$G_M(0)$	μ	$G_M(0)$	μ	$G_M(0)$	μ	$G_M(0)$	μ	$G_M(0)$	μ
N1	−2.272(199)	−0.757(67)	4.343(371)	1.447(125)	1.672(53)	0.412(13)	−2.199(173)	−0.701(56)	1.600(71)	0.389(18)
N2	−2.105(230)	−0.724(80)	3.747(466)	1.289(161)	1.609(47)	0.404(12)	−1.987(138)	−0.658(46)	1.567(47)	0.386(11)
N3	−2.516(627)	−0.897(223)	4.462(1.003)	1.591(358)	1.622(80)	0.410(20)	−1.863(129)	−0.621(44)	1.616(45)	0.400(11)
N4	−2.537(557)	−0.929(206)	5.098(1.050)	1.867(388)	1.635(74)	0.416(19)	−1.896(176)	−0.640(55)	1.621(50)	0.402(15)
Lin. Fit	−2.330(368)	−0.852(133)	4.295(700)	1.569(253)	1.602(58)	0.411(15)	−1.773(141)	−0.608(45)	1.625(47)	0.405(13)
χ^2/dof	–	0.44	–	0.993	–	0.186	–	0.129	–	0.149
p-Val	–	0.4	–	0.64	–	0.83	–	0.37	–	0.86
Quad. Fit	−2.891(736)	−1.073(269)	6.017(1.385)	2.220(505)	1.670(110)	0.425(29)	−1.903(276)	−0.639(88)	1.662(87)	0.413(24)
χ^2/dof	–	0.079	–	0.005	–	0.006	–	0.098	–	0.139
p-Val	–	0.7	–	0.78	–	0.94	–	0.95	–	0.71

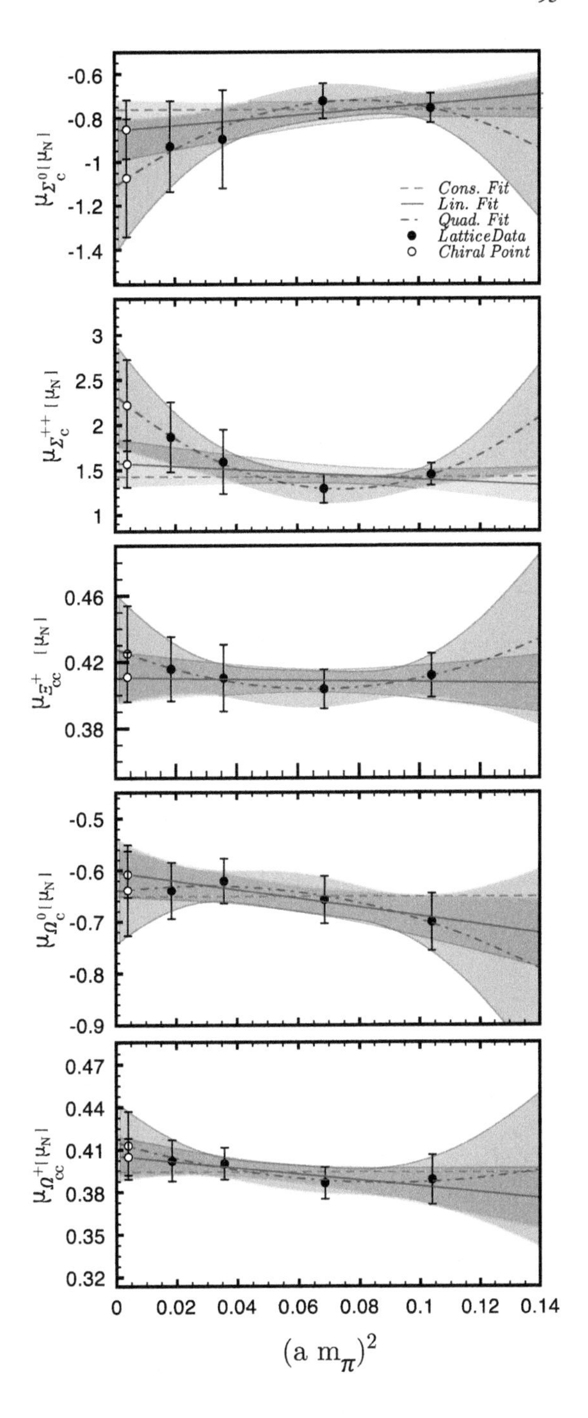

Fig. 5.17 Chiral extrapolations for the magnetic moments oftable 5.8 $\Sigma_c^{0,++}$, Ξ_{cc}^+, Ω_c^0 and Ω_{cc}^+. We show the fits to constant, linear and quadratic forms. Shaded regions are the maximally allowed error regions, which give the best fit to data. Figure taken from Ref. [16]

Table 5.8 Comparison of our results with various other models. All values are given in nuclear magnetons [μ_N]

	Our result		[23]	[30]	[25]	[26]	[27]	[28]	[29]	[30]	[31]
	Lin. fit	Quad. fit									
$\mu_{\Sigma_c^0}$	−0.875(103)	−1.117(198)	−1.78	−1.04	–	−1.043	−1.60	−1.391	−1.17	−1.015	−1.6(2)
$\mu_{\Sigma_c^{++}}$	1.499(202)	2.027(390)	3.07	1.76	–	1.679	2.20	2.44	2.18	2.279	2.1(3)
$\mu_{\Xi_{cc}^+}$	0.411(15)	0.425(29)	0.94	0.72	$0.785^{+0.050}_{-0.030}$	0.722	0.84	0.774	0.77	–	–
$\mu_{\Omega_c^0}$	−0.608(45)	−0.639(88)	−0.90	−0.85	–	−0.774	−0.90	−0.85	−0.92	−0.960	–
$\mu_{\Omega_{cc}^+}$	0.405(13)	0.413(24)	0.74	0.67	$0.635^{+0.012}_{-0.015}$	0.668	0.697	0.639	0.70	0.785	–

Table 5.9 Individual quark sector contributions to the magnetisation radii and the magnetic moments of the charmed baryons. Note that the numbers are given in units of fm^2 and nuclear magnetons respectively and independently from the electric charge of the individual quarks that compose the baryons

ID	$\Sigma_c^{0,++}$		$\Xi_{cc}^{+,++}$		Ω_c^0		Ω_{cc}^+	
	$\langle r_M^2\rangle_q$	$\langle r_M^2\rangle_Q$	$\langle r_M^2\rangle_q$	$\langle r_M^2\rangle_Q$	$\langle r_M^2\rangle_q$	$\langle r_M^2\rangle_Q$	$\langle r_M^2\rangle_q$	$\langle r_M^2\rangle_Q$
N1	0.444(55)	0.067(43)	0.471(38)	0.079(7)	0.424(53)	0.090(37)	0.405(49)	0.077(11)
N2	0.346(52)	0.063(25)	0.371(40)	0.078(8)	0.300(54)	0.064(21)	0.253(25)	0.073(9)
N3	0.394(69)	0.185(152)	0.473(58)	0.076(8)	0.405(44)	0.096(30)	0.367(40)	0.096(9)
N4	0.506(99)	0.141(165)	0.477(85)	0.085(8)	0.405(38)	0.053(18)	0.385(47)	0.088(10)
Lin. Fit	0.403(67)	0.098(80)	0.426(60)	0.082(6)	0.398(44)	0.056(19)	0.350(44)	0.095(9)
Quad. Fit	0.604(118)	0.236(183)	0.612(115)	0.089(11)	0.484(70)	0.054(38)	0.534(72)	0.101(16)
	$\Sigma_c^{0,++}$		$\Xi_{cc}^{+,++}$		Ω_c^0		Ω_{cc}^+	
ID	μ_q	μ_Q	μ_q	μ_Q	μ_q	μ_Q	μ_q	mu_Q
N1	2.178(188)	−0.080(16)	−0.474(34)	0.414(11)	2.080(155)	−0.071(17)	−0.402(32)	0.412(17)
N2	2.046(248)	−0.096(16)	−0.416(33)	0.421(9)	1.833(144)	−0.098(13)	−0.356(19)	0.411(11)
N3	2.427(572)	−0.115(26)	−0.434(52)	0.420(9)	1.785(144)	−0.088(10)	−0.370(27)	0.432(9)
N4	2.581(555)	−0.061(13)	−0.471(86)	0.432(11)	1.838(183)	−0.099(18)	−0.393(33)	0.436(15)
Lin. Fit	2.369(362)	−0.099(21)	−0.410(51)	0.430(8)	1.710(150)	−0.099(14)	−0.370(26)	0.441(12)
Quad. Fit	2.943(732)	−0.059(36)	−0.516(117)	0.433(16)	1.915(279)	−0.083(28)	−0.428(58)	0.453(22)

correctly determined, there is a large discrepancy among results. For all the baryons, the moments seem to be underestimated with respect to other methods.

Assessing the individual quark sectors is once again instructive to understand the internal dynamics of the baryons. In Table 5.9 we give the contributions of the quark sectors to the magnetic moments. A glance at the results show that, unlike the charge radii, features of the singly and doubly charmed baryons differ. In case of the singly charmed Σ_c and Ω_c baryons for instance, heavy quark contribution is much less compared to that of the light quark suggesting that the light quark give the dominant contribution to the magnetic moments of the baryons. Results are quite different when we examine the doubly charmed Ξ_{cc} and Ω_{cc} baryons: quark sector contributions are similar in magnitude.

While the magnitude of the contributions gives us an idea about the significance of the quark sector, their signs hold the information about the alignment of the spins of the quarks. We can infer by the opposite signs of the magnetic moments that the spins of the light and heavy quarks are anti-aligned in the baryon most of the time. Since the most significant contribution to the magnetic moments of the Σ_c and Ω_c comes from the doubly represented light sector, their spins are mainly determined by the light quarks. In general, doubly represented quarks tend to form spin-1 di-quark structures with their spins aligned with respect to each other. We observe this tendency in the doubly charmed baryons as well where the heavy-quark has a larger contribution to the total spin and magnetic moment.

Findings for the magnetic moments

- *General features*:
  - Magnetic moments of the charmed baryons are smaller in magnitude compared to their light counterparts.
  - Σ_c^{++} has the largest magnetic moment among the spin-1/2 baryons that we have studied.
- *Quark sectors*:
  - Contributions differ for singly- and doubly-charmed baryons. Light sector is dominant in the singly-charmed while the sectors are comparable in the doubly-charmed baryons.
  - Opposite signs of the quark magnetic moments suggest their spins are anti-aligned.

5.4 Spin-3/2 Baryons

We present the results of our simulations for the electromagnetic multipole form factors of the spin-3/2 Ω, Ω_c^*, Ω_{cc}^* and Ω_{ccc} baryons. Theoretical formalism has been outlined in Sect. 4.1.2.2. Spin-3/2 $\rightarrow$ spin-3/2 transition allows us to extract more information from the electromagnetic multipole form factors as compared to the spin$-1/2 \rightarrow$ spin$-1/2$ transition. In addition to the charge radii and magnetic moments we have access to the electric-quadrupole moments of the baryons as well. In the following sections we give the numerical values of the form factors, charge radii and moments and discuss their physical implications.

5.4.1 Electric Properties

5.4.1.1 Plateau Analysis

We extract the $E0$ and $E2$ multipole form factor values by searching for plateau regions of the ratio given in Eq. (4.44). Note that we were able to isolate a clear signal for the $E0$ form factors although the $E2$ form factors are much noisier compared to the dominant $E0$ form factor. Unfortunately the limited number of the gauge configurations we have for the the lightest quark mass ensemble N5, prevents us to reach a statistically significant value for the s-quark contributions of the $E2$ moments since its signal is too noisy.

There are some differences to note in contrast to the spin-1/2 analysis: First of all, all the measurements are performed on the ensemble N5. The results we quote are not extrapolated to the physical light-quark mass since the mass of the light quarks on the gauge configurations of this ensemble is almost physical. We regard the results as final. Secondly, we calculate the form factors only for the lowest allowed lattice momentum transfer. We will show that this approach leads to consistent results with the dipole-form fit that we have employed for the spin-1/2 case. In Figs. 5.18 and 5.19 we show the correlation-function ratios of the $E0$, and $E2$ form factors for the strange and charm quark sectors and the plateau fit extracted values at the lowest allowed three-momentum transfer ($\mathbf{q}^2$=0.183 GeV^2) are given in Table 5.10. Note that $E0$ form factor reduces to the electric charge of the baryon as usual and the

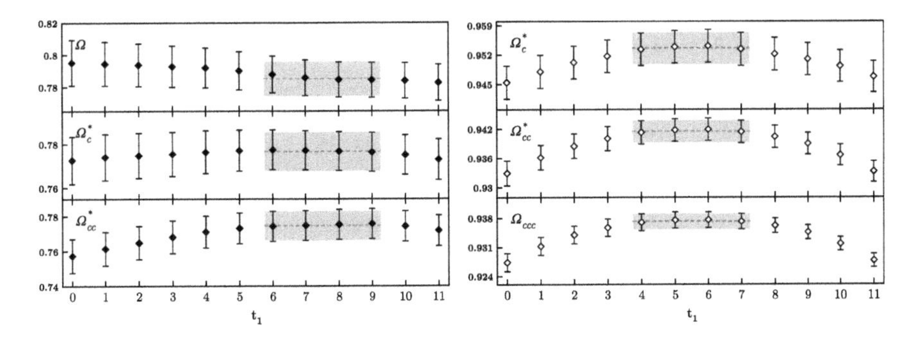

Fig. 5.18 Strange (filled) and charm quark (empty) contributions to the $E0$ form factor at the lowest allowed three-momentum transfer ($\mathbf{q}^2$=0.183 GeV^2). Contributions are shown for a single quark and normalised to unit charge. The fit regions are $t_1 = [4, 7]$ for the charm sector and $t_1 = [6, 9]$ for the strange sector. Figure taken from Ref. [32]

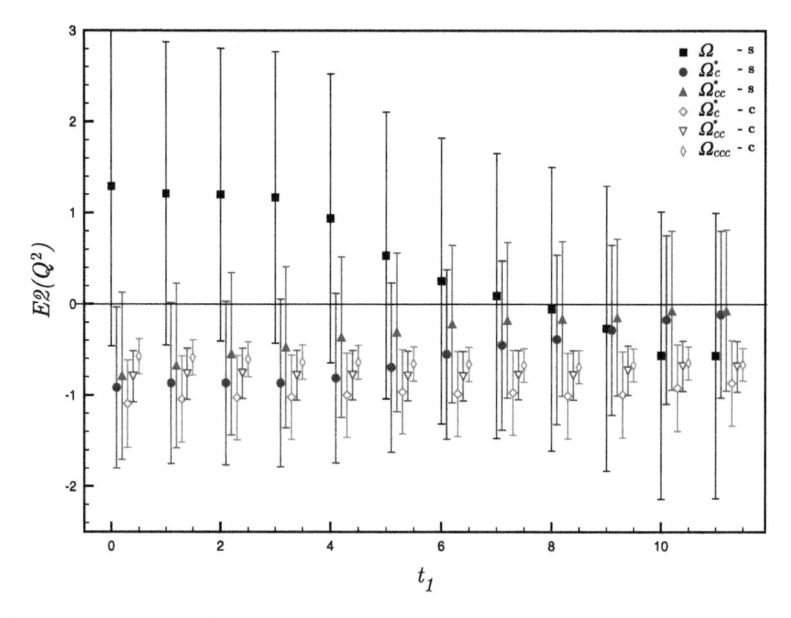

Fig. 5.19 Same as Fig. 5.18 but for the $E2$ form factor. Fit region is $t_1 = [4, 7]$ for all cases. Data points are slightly shifted for a clear view. Figure taken from Ref. [32]

$E2$ form factor cannot be directly obtained at zero momentum transfer due to its definition in Eq. (4.46).

Excited states are of concern in spin-3/2 analysis as well, however, we have shown that a separation of $\sim$1.09 fm ($12a$) is sufficient to keep the contamination under control in the spin-1/2 case. Despite that, we further check possible contaminations by extracting the $E0$ form factor via summed operator insertions method. In Fig. 5.20, $E0$ form factors of the Ω^*_{cc} and Ω_{ccc} baryons are given. We expect the excited state contaminations to be severest for those baryons relying on the heavy-quark spin symmetry, which suggests that the energy gap between the ground and excited states decrease as the mass of the quarks increase. Yet, the values extracted by a plateau and SOI method agree with each other very well indicating that the contamination is under control.

Table 5.10 Values of the $E0(Q^2)$ and $E2(Q^2)$ form factors at $\mathbf{q}^2$=0.183 GeV^2 for Ω, Ω^*_c, Ω^*_{cc} and Ω_{ccc}. Results are given in lattice units for a single quark and normalised to unit charge

	$E0^s(Q^2)$	$E0^c(Q^2)$	$E2^s(Q^2)$	$E2^c(Q^2)$
Ω	0.789(12)	–	−0.228(773)	–
Ω^*_c	0.778(9)	0.954(4)	−0.630(915)	−0.979(456)
Ω^*_{cc}	0.775(8)	0.942(2)	−0.280(852)	−0.787(266)
Ω_{ccc}	–	0.937(2)	–	−0.655(182)

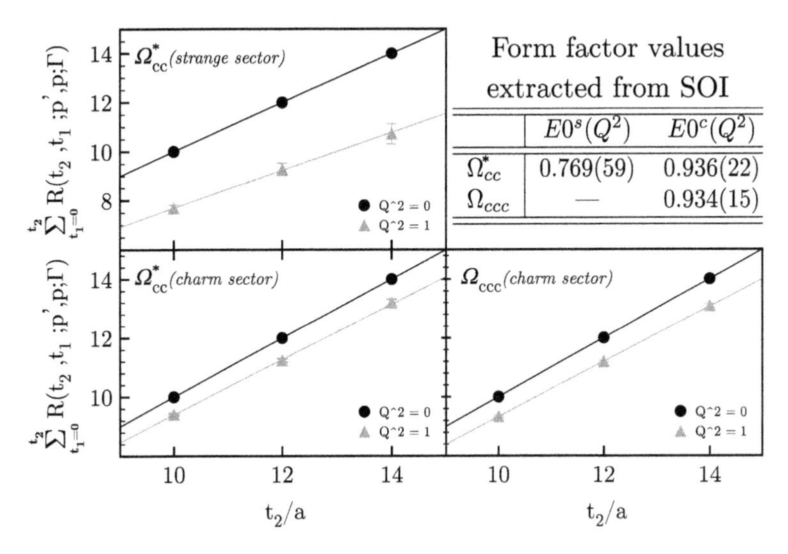

Fig. 5.20 $E0$ form factors of Ω^*_{cc} and Ω_{ccc} baryons for two momentum insertions as obtained from summed operator insertions method. Results are for 43 measurements. Values given in the table are for $Q^2 = 1$ and should be compared to Table 5.10. Figure taken from Ref. [32]

Table 5.11 Electric charge radii of the Ω, Ω^*_c, Ω^*_{cc} and Ω_{ccc}. Results are given in fm^2. Quark sector contributions are for single quark and normalised to unit charge. Electric charge radii of spin-1/2 baryons are estimated through form factor fits as in Sect. 5.3. Total electric charge radius of the spin-1/2 Ω_c is estimated by the Eq. (5.18) since its electric form factor vanishes due to its zero electric charge

	$\langle r_E^2\rangle_s$	$\langle r_E^2\rangle_c$	$\langle r_E^2\rangle$
Ω_c	0.329(25)	0.064(11)	−0.177(18)
Ω_{cc}	0.313(16)	0.073(4)	0.026(4)
Ω	0.326(21)	–	−0.326(21)
Ω^*_c	0.345(17)	0.062(5)	−0.189(12)
Ω^*_{cc}	0.348(16)	0.078(3)	−0.012(6)
Ω_{ccc}	–	0.084(3)	0.168(5)

Table 5.12 Electric charge radius of Ω^*_{cc} and Ω_{ccc} baryons extracted by the plateau and the SOI method. Results are compared for 43 measurements. Charge radii are given in fm^2

	Ω^*_{cc}		Ω_{ccc}	
	Plateau	SOI	Plateau	SOI
$\langle r_E^2\rangle_s$	0.332(43)	0.359(113)	–	–
$\langle r_E^2\rangle_c$	0.079(8)	0.086(31)	0.085(5)	0.089(21)
$\langle r_E^2\rangle$	−0.005(16)	−0.006(53)	0.170(9)	0.179(42)

5.4.1.2 Electric Charge Radii

Since we perform our simulations with a single value of the finite momentum transfer, a dipole-form fit is not possible. However, we assume that a dipole form ansätz holds and extract the charge radii using the expression,

$$\frac{\langle r_E^2 \rangle}{G_{E0}(0)} = \frac{12}{Q_{min}^2}\left(\sqrt{\frac{G_{E0}(0)}{G_{E0}(Q_{min}^2)}} - 1\right), \tag{5.22}$$

which can be readily derived by inserting Eq. (5.12) into Eq. (5.11). We compile the electric charge radii of the baryons as well as the quark sector contributions extracted by the above expression in Table 5.11. In Fig. 5.20, we have shown the form factor values extracted by the SOI method in comparison to plateau method results of Table 5.10 to make sure that the excited state contamination is under control. We utilise those values to give a comparison of the electric charge radii estimated by using the form factor values extracted via the plateau and SOI method in Table 5.12. Agreement between the results shows that we have good plateau signals which isolate the ground state effectively.

For the ease of discussion, we plot the data of Table 5.11 in Fig. 5.21. Contribution of the s-quark to the electric charge radii in all baryons, shown in Fig. 5.21a, appears to be similar to each other, implying that it is almost independent of the quark-flavor composition of the baryon. Moving on to the c-quark contributions, we see that although the contributions illustrated in Fig. 5.21b seems to increase slightly with the increasing number of the valence c-quarks in the baryon, smallness of the scale makes this change negligible compared to the s-quark sector.

Effects of the spin-vector alignment of the quark to the charge radii are observed when we compare the spin-1/2 and spin-3/2 sectors to each other. Apparent agreement between the contributions in spin-1/2 and spin-3/2 sectors shows that the effect of the spin on the s- and c-quarks is almost non-existent. We can form ratios of the individual quark-flavor contributions in spin-1/2 to that in the spin-3/2 sector as $\langle r_E^2 \rangle_B^q / \langle r_E^2 \rangle_{B*}^q$, in order to study the spin-effect more systematically. In Fig. 5.21c we show the ratios for the s- and c-quarks in singly- and doubly-charmed baryons. In the singly-charmed Ω_c baryon we observe that the s- and c-quark charge distributions are insensitive to the spin-flip of the c-quark while a deviation from one is an indication of an increase in their contributions in case of the doubly-charmed Ω_{cc} baryon.

Total electric charge radii of the baryons given in the final column of Table 5.11 are evaluated via the Eq. (5.18) and illustrated in Fig. 5.21d. In magnitude, Ω baryon has the largest electric charge radius among all baryons we study. We should note that the electric charge radius of the Ω baryon, $\langle r_E^2 \rangle_{\Omega^-} = -0.326(21)$ fm^2, is in quite good agreement with the previous lattice determinations [15, 19]. Ω_c, Ω_c^* and Ω_{ccc} seem to have similar charge radii while the charge radii of the Ω_{cc} and Ω_{cc}^* almost vanish.

We can make a naive assumption based on the similarity of the quark contributions to the charge radii and consider the quark sector contributions in the spin-3/2 sector

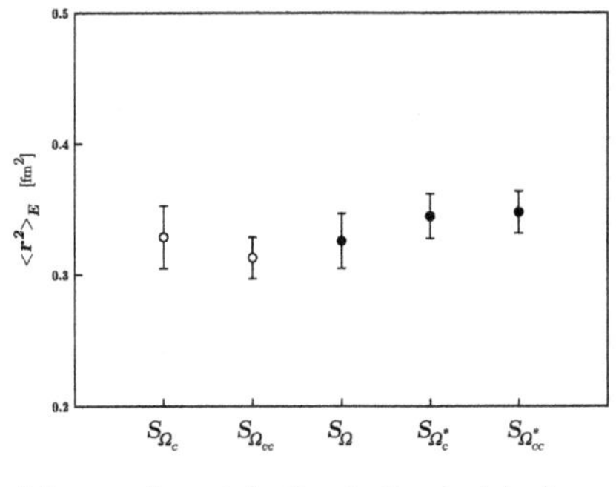

(a) s-quark contribution to the electric charge radii.

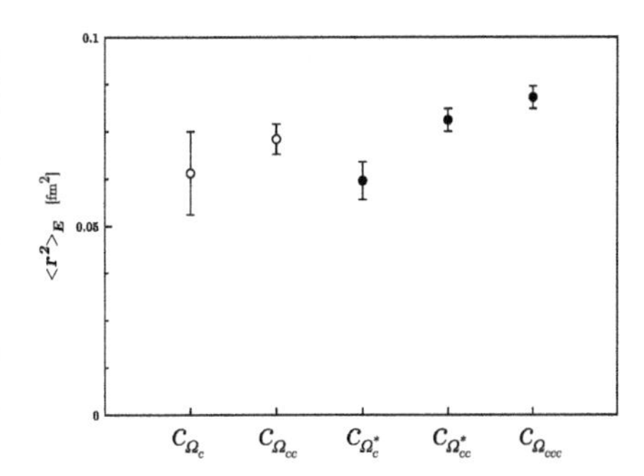

(b) c-quark contribution to the electric charge radii.

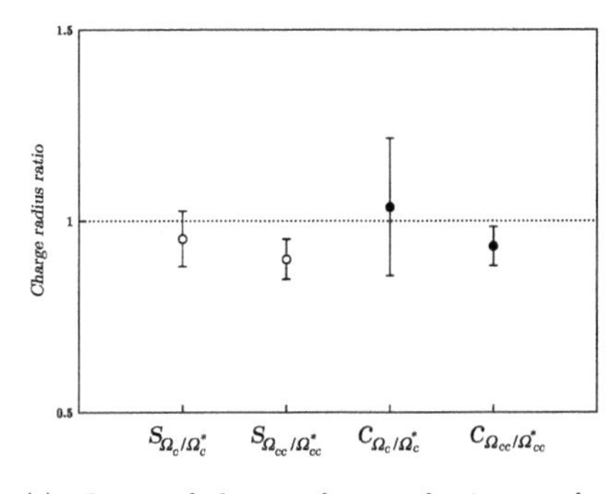

(c) Ratio of the quark contribution to the electric charge radii.

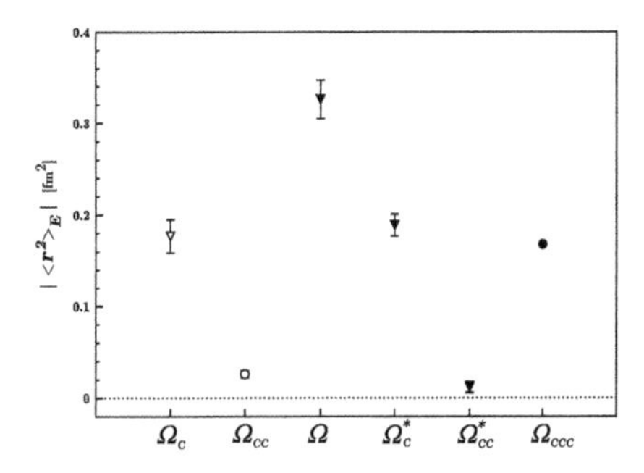

(d) The electric charge radii of the baryons that we consider.

Fig. 5.21 Individual quark sector contributions to the electric charge radii, ratios of those contributions and the total electric charge radii of the spin-1/2 Ω_c, Ω_{cc} and spin-3/2 Ω, Ω_c^*, Ω_{cc}^*, Ω_{ccc} baryons. q_{B/B^*} in (**c**) is a shorthand notation for the ratio, $\langle r_E^2 \rangle_B^q / \langle r_E^2 \rangle_{B^*}^q$. Absolute values are shown for a better comparison in (**d**). Data points denoted by a triangle indicate a negative value. Figures taken from Ref. [32]

to be same so that, $\langle r_E^2 \rangle_\Omega^s = \langle r_E^2 \rangle_{\Omega_c^*}^s = \langle r_E^2 \rangle_{\Omega_{cc}^*}^s = R_s^2$ and $\langle r_E^2 \rangle_{\Omega_c^*}^c = \langle r_E^2 \rangle_{\Omega_{cc}^*}^c = \langle r_E^2 \rangle_{\Omega_{ccc}}^c = R_c^2$. With the help of the Eq. (5.18) we can relate the electric charge radii of the spin-3/2 baryons to each other as, $\left(\langle r_E^2 \rangle_{\Omega_c^*} + \langle r_E^2 \rangle_{\Omega_{ccc}} \right) /2 = \langle r_E^2 \rangle_{\Omega_{cc}^*}$. Inputting our results, charge radius of Ω_{cc}^* evaluates to $\left(\langle r_E^2 \rangle_{\Omega_c^*} + \langle r_E^2 \rangle_{\Omega_{ccc}} \right) /2 = -0.011(8)$, which we compare to the computed charge radius of Ω_{cc}^*, $\langle r_E^2 \rangle_{\Omega_{cc}^*} = -0.012(6)$. Agreement shows that this relation holds nicely which implies that the contribution to the charge radii from each flavor is similar for all baryons we consider here and their radii differ due to different quark compositions they have.

In order to get a better idea about the interplay of the quark sectors within the light and heavy baryons, it is instructive to compare the behaviour of the s-quark contributions to the electric charge radii of the Ω_c^* (ssc), Ω_{cc}^* (scc) to that of Ξ^* (ssu), Σ^* (suu) baryons. We quote the values of Ref. [15] in which the authors have calculated the same observables for decuplet baryons. A comparison of s-quark

Table 5.13 Single strange quark contribution to the electric charge radii of the Ω_c^* and Ω_{cc}^* (normalised to unit charge) in comparison to that of the decuplet Ξ^* and Σ^*. Decuplet values are taken from tables XI. and XII. of Ref. [15]

	$S_{\Omega_c^*}$	S_{Ξ^*}	$S_{\Omega_{cc}^*}$	S_{Σ^*}
$\langle r_E^2 \rangle$ [fm^2]	0.345(17)	0.308(17)	0.348(16)	0.321(22)

Table 5.14 Electric charge radii of Ω_c extracted by a dipole fit to the electric form factor and by use of Eq. (5.22)

	Dipole fit	Equation (5.22)
$\langle r_E^2 \rangle_s$ [fm^2]	0.329(25)	0.338(26)
$\langle r_E^2 \rangle_c$ [fm^2]	0.064(11)	0.067(11)
$\langle r_E^2 \rangle$ [fm^2]	−0.177(18)	−0.180(20)

electric charge radii in Ω_c^* - Ξ^* in Table 5.13 reveals the effect of changing the single u-quark by a c-quark: When the singly represented quark is heavier, the s-quark charge radius increases. In case of the Ω_{cc}^* - Σ^* baryons, the doubly represented light quarks are changed to c-quarks. While the current precision does not allow a clear conclusion, such a comparison again suggests a slight increase in the charge radius.

Note on systematic errors: Unlike the analysis of the spin-1/2 baryons, where we have performed dipole-form fits to a range of form factor values corresponding to different momenta to extract the charge radii via Eq. (5.13), we estimate the charge radii of the spin-3/2 baryons by evaluating the Eq. (5.22). That expression greatly simplifies the simulations and allows us to extract the values more precisely since we use only two sets of momentum values, namely $[Q^2 = 0, Q^2 = 1]$, and avoid the possible statistical fluctuation due to higher momenta. We derived Eq. (5.22) by assuming the data is described by a dipole form, however, we need to check whether our assumption holds or not. As an illustrative example, we compare the charge radii evaluated via the procedure we have followed for the spin-1/2 baryons to the values extracted via Eq. (5.22) for the spin-1/2 Ω_c baryon. Results are given in Table 5.14 where the middle column holds the dipole-fit extracted values and the charge radii estimated via Eq. (5.22) are quoted in the rightmost column. Considering the current precision, we conclude that both approaches agree and no systematic error arises due to change of analysis.

Findings for the electric charge radii

- *General features*:
  - Ω^- baryon has the largest charge radius in magnitude among the baryons that we study.
  - Its value $\langle r_E^2 \rangle_{\Omega^-} = -0.326(21)$ fm^2 is in good agreement with other determinations [15, 19].
  - Ω_c^* and Ω_{ccc} have similar charge radii while Ω_{cc}^* has almost vanishing radius.

- *Quark sectors*:
  - Strange and charm quark charge radii are insensitive to the baryon quark-flavor composition.
  - For singly charmed baryons, s- and c-quark charge radii are not affected by the spin-flip whereas the charge radii of doubly charmed baryons increase.

5.4.1.3 Electric-Quadrupole Moment

The electric-quadrupole moment gives a measure of the deviation of the electric charge from a spherically symmetric distribution as we have discussed in Sect. 5.2. It also hints to the tensor force interactions. Similar to the analysis of the $E0$ form factor, we estimate the $E2$ form factor by the plateau approach. We compute and extract the s- and c-quark sector contributions individually. We have shown the $E2$ form factors in Fig. 5.19 and given the numerical values in lattice units in Table 5.10. In Table 5.15, we quote the values in physical units of $[e/m^2]$. Note that the results are obtained at the smallest three-momentum value of $\mathbf{q}^2 = 0.183$ GeV^2. Rather than a precise determination of the quadrupole moments we are interested in distinguishing their signs so that we can estimate the deformation of the electric charge distribution.

The evident poor signal in Fig. 5.19 prevents us from isolating the sign of the quadrupole moments of the Ω and Ω_c^* baryons. Better statistical precision of the heavy quarks, however, weights in when we consider the heavier Ω_{cc}^* and Ω_{ccc} baryons and we are able to isolate the sign of their moments. Ω_{cc}^{*+} and Ω_{ccc}^{++} have negative $E2$ moments thus their charge distributions deform to an *oblate* shape.

Table 5.15 $E2(Q^2)$ results for Ω, Ω_c^*, Ω_{cc}^* and Ω_{ccc} at $\mathbf{q}^2 = 0.183$ GeV^2. Values are given in units of $[e/m^2]$. Quark sector contributions are for single quark and normalised to unit charge. Last column is calculated by the Eq. (5.18)

	$E_2(Q^2)_s$	$E_2(Q^2)_c$	$E_2(Q^2)$
Ω	−0.337(1.142)	–	0.337(1.142)
Ω_c^*	−0.371(539)	−0.577(269)	−0.137(352)
Ω_{cc}^*	−0.091(277)	−0.255(87)	−0.310(128)
Ω_{ccc}	–	−0.136(38)	−0.273(76)

Table 5.16 Values of $M1(Q^2)$ form factor at $\mathbf{q}^2$=0.183 GeV^2 for Ω, Ω_c^*, Ω_{cc}^* and Ω_{ccc}. Results are given in lattice units for single quark and normalised to unit charge

	$M1^s(Q^2)$	$M1^c(Q^2)$
Ω	2.307(94)	–
Ω_c^*	3.413(96)	1.032(25)
Ω_{cc}^*	4.442(110)	1.349(16)
Ω_{ccc}	–	1.609(12)

Findings for the quadrupole moments

Ω_{cc}^{*+} and Ω_{ccc}^{++} have oblate charge distribution.

5.4.2 Magnetic Properties

5.4.2.1 Plateau Analysis

In the magnetic sector we extract the $M1$ multipole form factor only. Data for the $M3$ form factor is too noisy to isolate a statistically significant value. We omit the $M3$ form factor in this work since the limited number of gauge configurations prevents us from increasing the quality of the signal. Similar to the electric sector we perform simulations only on the ensemble N5 and with the lowest allowed lattice three-momentum. In Fig. 5.22, we show the ratio in Eq. (4.44) for the $M1$ form factors of the strange and charm quark sectors as well as the identified plateau regions. Extracted values at the lowest allowed three-momentum transfer ($\mathbf{q}^2$=0.183 GeV^2) are compiled into Table 5.16.

5.4.2.2 Magnetic Moments

We need the zero-momentum value, $G_{M1}(0)$, of the magnetic-dipole form factor in order to calculate the magnetic moments of the baryons. We calculate the form factor value via the Eq. (5.14) and simply evaluate the Eq. (5.15) to estimate the magnetic moments. Our numerical values are listed in Table 5.17 and illustrated in Fig. 5.23.

Based on the data collected from ensembles N1–N4, we have concluded in Sect. 5.3.2 that a single quark's contribution to the magnetic moment increases significantly when it is doubly represented in spin-1/2 baryons. Results for the Ω_c and Ω_{cc} baryons from the ensemble N5 confirms our conclusion. When we contrast quark sector contributions of the spin-1/2 Ω_c and Ω_{cc} baryons to that of the spin-3/2 Ω_c^* and

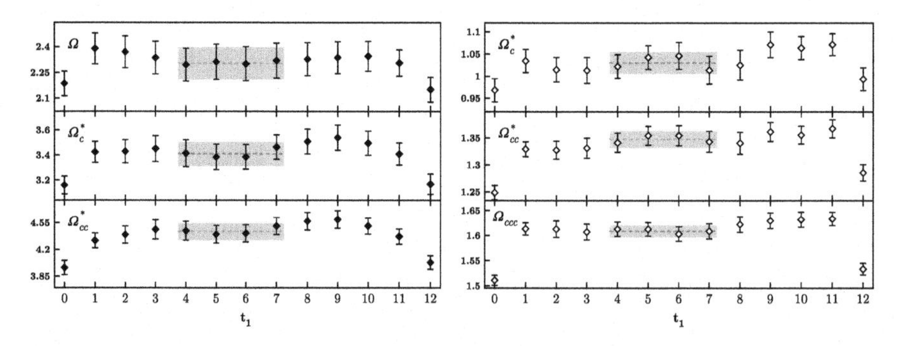

Fig. 5.22 Same as Fig. 5.18 but for the $M1$ form factor. Fit region is $t_1 = [4, 7]$ for all cases. Figure taken from Ref. [32]

Table 5.17 Magnetic moments of Ω, Ω_c^*, Ω_{cc}^* and Ω_{ccc}. Results are given in units of nuclear magnetons, μ_N. Quark sector contributions are for single quark and normalised to unit charge

	μ_s	μ_c	μ
Ω_c	0.979(47)	−0.092(6)	−0.688(31)
Ω_{cc}	−0.402(17)	0.216(3)	0.403(7)
Ω	1.533(55)	–	−1.533(55)
Ω_c^*	1.453(36)	0.358(8)	−0.730(23)
Ω_{cc}^*	1.408(29)	0.352(4)	0.000(10)
Ω_{ccc}	–	0.338(2)	0.676(5)

Table 5.18 Strange quark contributions to the magnetic moments of the Ω_c^*, Ω_{cc}^*, Σ^* and Ξ^*. Decuplet baryon results are calculated in Ref. [15] on a quenched configuration with $m_\pi = 300$ MeV. All contributions are for a single strange quark of unit charge

	$s_{\Omega_c^*}$	s_{Ξ^*}	$s_{\Omega_{cc}^*}$	s_{Σ^*}
μ [μ_N]	1.453(36)	1.725(77)	1.408(29)	1.750(10)

Ω_{cc}^* baryons, we see that a sign change is evident due to the spin flip. Contributions in the spin-3/2 sector are larger compared to the spin-1/2, which can be understood by the change in the configuration of the baryons. In order to compose a spin-1/2 baryon, one of the quark sectors should be anti-aligned with the spin vector of the baryon causing an overall decrease in the contributions unlike the spin-3/2 baryons, where the spin vectors of all the quarks are aligned both with themselves and the baryon itself which enhances the contributions. Within the spin-3/2 baryons on the other hand, s-quark contributions have a slight tendency to decrease with the decreasing number of valence s-quarks. Contributions of the c-quark, however, tend to decrease as the number of c-quarks increase.

We may investigate the differences occurring due to changes in the quark flavor composition via comparing the light sector to the heavy sector in a similar fashion to that in the electric sector. Considering the same baryons, Ω_c^* (ssc) and Ω_{cc}^* (scc) in comparison to Ξ^* (ssu) and Σ^* (suu), we contrast the magnetic moment of the s-quark to search for the effects of changing a light quark by a charm quark. Magnetic moments of light decuplet baryons have been calculated in Ref. [15] with quenched lattice QCD. In Table 5.18, we compile our results along with the results of Ref. [15]. Although a quantitative comparison is impractical since the work in Ref. [15] has been performed on quenched lattices with much heavier pion mass, we can assert a qualitative comparison as follows: The s-quark contributions to the magnetic moments of the charmed and light decuplet baryons are different. Charmed baryons have smaller magnetic moments than light baryons.

We can study the effect of the quark spin configurations on the quark magnetic moments further by forming the ratios of the quark-sector contributions to spin-1/2 and spin-3/2 baryons, $\mu_B^q/\mu_{B^*}^q$, where q is the quark flavor and B is the baryon. We compile the numerical values in Table 5.19 together with the octet-decuplet ratios

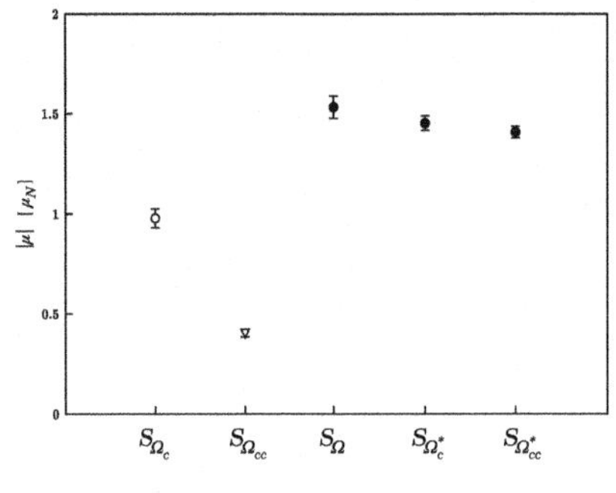

(a) s-quark contributions to the magnetic moments.

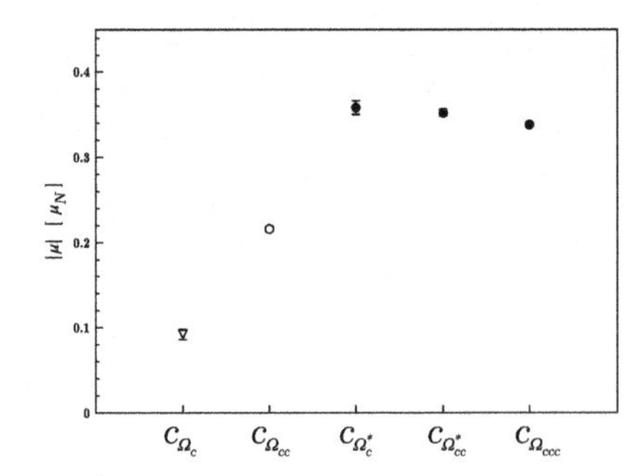

(b) c-quark contributions to the magnetic moments.

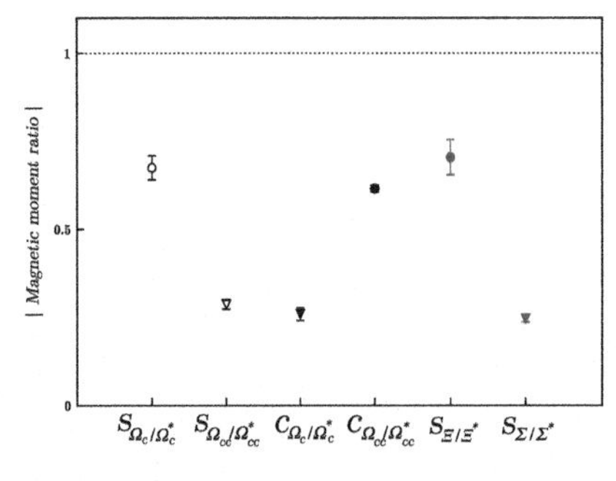

(c) Ratios of the quark contributions to the magnetic moments.

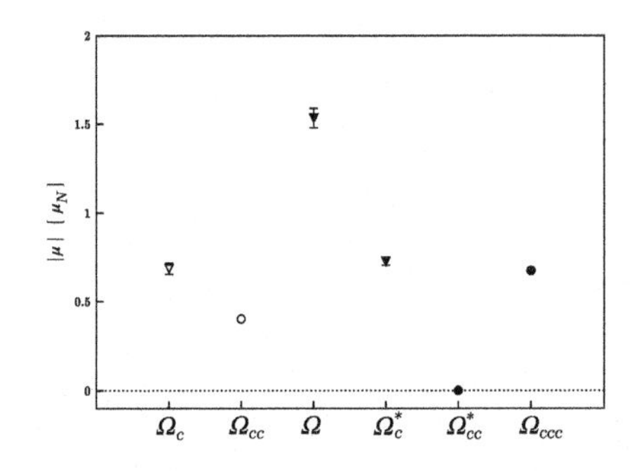

(d) Total magnetic moments of the baryons that we consider.

Fig. 5.23 Individual quark sector contributions to the magnetic moments, ratios of those contributions and the total magnetic moments of the spin-1/2 Ω_c, Ω_{cc} and spin-3/2 Ω, Ω_c^*, Ω_{cc}^*, Ω_{ccc} baryons. Absolute values are shown for a better comparison. Data points denoted by a triangle indicate negative values. Rightmost blue data points in (**c**) are octet/decuplet ratios calculated using the $m_\pi = 300$ MeV quenched simulation results of Refs. [14, 15] and q_{B/B^*} is a shorthand notation for $\mu_B^q/\mu_{B^*}^q$ where q is the quark flavor and B is the baryon. Figures taken from Ref. [32]

extracted from Refs. [14, 15] and illustrate the data in Fig. 5.23c for the ease of discussion. It is evident that the contributions from both of the quark sectors are enhanced in spin-3/2 baryons. Notice that the strange quark sector in the Ω_c and Ω_c^* and the charm quark sector in Ω_{cc} and Ω_{cc}^* baryons are doubly represented and the difference is that a strange quark is exchanged by a charm quark or vice versa. A comparison of the ratios, S_{Ω_c/Ω_c^*} and $C_{\Omega_{cc}/\Omega_{cc}^*}$, would reveal the effect of such a quark flavor exchange. Same exercise can be done for the singly strange and singly charmed baryon ratios also. We observe that results of these ratios are respectively consistent with each other suggesting that the flavor of the quark has almost no role in the difference between the spin-1/2 and the spin-3/2 baryons.

Including the magnetic moment of the s-quark sectors of the octet and decuplet baryons, namely for the Σ, Ξ, Σ^* and Ξ^*, we may enhance our understanding of

Table 5.19 Ratios of the quark magnetic moment contributions in $1/2^+/3/2^+$. Octet/decuplet ratios are extracted from the numerical results available in the Refs. [14, 15]. All values are ratios of a single quark contribution of unit charge

	S_{Ω_c/Ω_c^*}	$C_{\Omega_{cc}/\Omega_{cc}^*}$	S_{Ξ/Ξ^*}
$\lvert\mu_B^q/\mu_{B^*}^q\rvert$	0.674(34)	0.615(10)	0.703(50)
	$S_{\Omega_{cc}/\Omega_{cc}^*}$	C_{Ω_c/Ω_c^*}	S_{Σ/Σ^*}
$\lvert\mu_B^q/\mu_{B^*}^q\rvert$	0.286(13)	0.258(18)	0.245(10)

the behaviour of strange sector with respect to its environment. In case of the doubly strange baryons the ratios S_{Ω_c/Ω_c^*} and S_{Ξ/Ξ^*} agree with each other within their one-sigma errors suggesting that the strange quark is insensitive to its accompanying quark—be it a light quark or a heavy charm quark. However, we see a slight discrepancy in singly-strange baryons when we compare the $S_{\Omega_{cc}/\Omega_{cc}^*}$ ratio to S_{Σ/Σ^*}. The smaller value of the S_{Σ/Σ^*} ratio suggests that going from spin-1/2 to spin-3/2 affects the s-quark more when its accompanying quarks are light flavored. Adding the C_{Ω_c/Ω_c^*} ratio into the comparison of the contributions of the singly-represented quark sectors, we see that the effect of the environment is less pronounced for the charm quark.

We combine the magnetic moments of the individual quark sectors via Eq. (5.18) and estimate the total magnetic moments of the baryons. The last column of Table 5.17 holds the total magnetic moments and we illustrate a comparison in Fig. 5.23d. We find the magnetic moment of the Ω^- baryon to be $\mu_{\Omega^-} = -1.533 \pm 0.055\ \mu_N$, smaller compared to the current experimental value of, $\mu_{\Omega^-}^{exp} = -2.02 \pm 0.05\ \mu_N$ [20]. One of the reasons for the discrepancy is the mass of the Ω baryon as determined in our simulations since magnetic moments are sensitive to the mass of the baryon by definition. Our mass value, $m_\Omega = 1.790(17)$ GeV, differs by 7% from the experimental mass $m_\Omega = 1.673(29)$ GeV. Another plausible possibility is that different simulation methods might lead to different results. For instance, a determination in Ref. [33] by a background field method finds $\mu_{\Omega^-} = -1.93 \pm 0.08\ \mu_N$ on $m_\pi = 366$ MeV lattices while our result is in agreement with other determinations using the same method as ours: the quenched calculation of Boinepalli et.al [15] results in $\mu_{\Omega^-} = -1.697 \pm 0.065\ \mu_N$ and the Alexandrou et.al finds $\mu_{\Omega^-} = -1.875 \pm 0.399\ \mu_N$ [19] via an extrapolation to the physical point.

Moving on to the charmed baryons, we see that the magnetic moments of the spin-1/2 Ω_c and spin-3/2 Ω_c^* baryons are almost the same, which indicates that the spin flip of the charm quark has minimal effect, in agreement with the heavy-quark spin symmetry expectations. From a quark-model perspective we would expect the magnetic moments of the Ω_c (Ω_{cc}) and Ω_c^* (Ω_{cc}^*) to be similar to each other and we see that such an expectation holds for the Ω_c and Ω_c^* while there is a striking discrepancy for the Ω_{cc} and Ω_{cc}^* baryon where the latter has a vanishing magnetic moment. Ω_c and Ω_c^* has the same quark content but their spin configurations differ from each other in a way that the spin vector of the single c-quark is anti-aligned with that of the (ss) component in Ω_c in contrast to the Ω_c^* where the spin vectors of all the quarks are aligned. When the quark sectors are combined relative to their electric charges, they add constructively for the Ω_c baryon but destructively for the Ω_c^*. Although the behaviour is different, combination of the sectors happen in a balanced way leading to similar magnetic moments for the Ω_c and the Ω_c^*. The balance is broken in case of the doubly-charmed Ω_{cc} and Ω_{cc}^* and the interplay of the electric charge and the number of the valence quarks ends up in a significant difference of the magnetic moments of the baryons.

Findings for the magnetic moments

- *General features*:
  - Magnetic moment of the Ω^- baryon is found to be, $\mu_{\Omega^-} = -1.533 \pm 0.055$ μ_N.
  - Ω_c, Ω_c^* have similar magnetic moments in magnitude.
  - Magnetic moment of the Ω_{cc}^* vanishes unlike the Ω_{cc}.
  - As compared to the decuplet sector, strange-quark contributions to the magnetic moments of spin-3/2 charmed baryons are smaller.
- *Quark sectors*:
  - Contributions amongst the spin-3/2 baryons are similar to each other, consonant with the quark-model expectations.
  - Magnetic moments of the strange and charm quarks in spin-3/2 charmed baryons are larger than spin-1/2 baryons having a similar quark-flavor composition.

5.5 Systematic Errors

Most of the estimates of systematic errors below are covered in their respective discussions already, however, we find it convenient to summarise the systematic uncertainties that we have encountered in the progress of this work, along with our conclusions, in one place for better accessibility.

5.5.1 Excited-State Contamination

Excited state contamination is one of the notorious systematic effects one should be worried of in matrix element calculations. We have covered its origin and the methods we use to determine the excited state effects in Sect. 5.2. Sections 5.3.1.1, 5.3.2.1, 5.4.1.1 and 5.4.2.1 contain our through checks of the excited state contamination by varying the source-sink time seperation and comparing the plateau extracted values with that of the phenomenological fit or SOI methods.

Figures 5.8 and 5.14 show the time dependence of the Ξ_{cc} form factors in case of the $t_2 = 12a$ and the longer $t_2 = 14a$ seperation. Behaviour of the data is similar

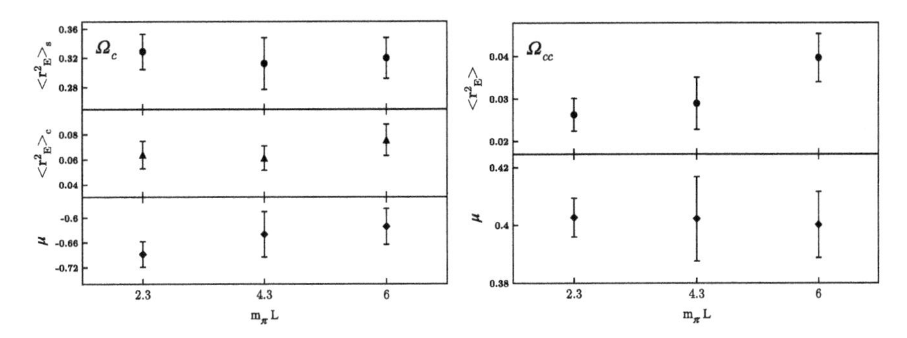

Fig. 5.24 Electric charge radius and magnetic moment of Ω_c and Ω_{cc} baryons on different $m_\pi L$ values. Since the Ω_c baryon has zero net electric charge we give the strange and charm quark charge radii. Figure taken from Ref. [32]

in both cases and plateau regions can be identified in the same fit windows. A comparison of the $t_2 = 12a$ plateau data with the data extracted via a phenomenological fit is depicted in Table 5.2 and plotted in Fig. 5.9 for the Ξ_{cc} baryon. We show the values extracted by a SOI method in Fig. 5.9 also. Form factor values and dipole fits of all three methods agree with each other within statistical errors with an increased uncertainty on phenomenological and SOI values.

We perform a plateau-SOI comparison in case of the spin-3/2 Ω^*_{cc} and Ω_{ccc} baryons, which would exhibit the severest excited state contamination among the spin-3/2 baryons. In Fig. 5.20 we show the $E0$ form factors of those baryons as extracted by a SOI method. Results are in good agreement with the plateau values in Table 5.10, e.g. $\left(E0^c(Q^2)\right)^{\Omega_{ccc}}_{\text{plateau}} = 0.937(2)$ compared to $\left(E0^c(Q^2)\right)^{\Omega_{ccc}}_{\text{SOI}} = 0.934(15)$. Furthermore, electric charge radii extracted from plateau and SOI values agree, as well. Comparison is given in Table 5.12. For instance, a plateau analysis yields $\left(\langle r_E^2\rangle^s\right)^{\Omega^*_{cc}}_{\text{plateau}} = 0.332(43)$ fm^2, in comparison to the SOI one, $\left(\langle r_E^2\rangle^s\right)^{\Omega^*_{cc}}_{\text{SOI}} = 0.359(113)$ fm^2.

We conclude that a source-sink seperation of $t_2 = 12a \approx 1.09$ fm is adequate to suppress the excited state contributions in case of spin-1/2 and spin-3/2 form factors and observables.

5.5.2 *Finite-Volume Effects*

Finite-volume corrections has an exponential dependency to the typical mass and spatial extent of the system as, e^{-ML} [34]. Studies put an empirical bound of $m_\pi L \geq 4$ to ensure the finite-size effects are negligible. Below this value, corrections are expected to be significant. The ensemble we use for spin-3/2 analysis, namely the N5, corresponds to $m_\pi L = 2.3$, which lies below the empirical one, thus, raises concerns about the finite-size effects. In order to fully account for this systematic effect, cal-

Table 5.20 Electric charge radius and magnetic moment of Ω_c and Ω_{cc} baryons on different $m_\pi L$ values

	$m_\pi L$	2.3	4.3	6
Ω_c	$\langle r_E^2 \rangle_s$ [fm^2]	0.329(24)	0.313(36)	0.320(28)
	$\langle r_E^2 \rangle_c$ [fm^2]	0.064(11)	0.061(10)	0.076(13)
	μ [μ_N]	−0.688(31)	−0.640(55)	−0.621(44)
Ω_{cc}	$\langle r_E^2 \rangle$ [fm^2]	0.026(4)	0.029(6)	0.040(6)
	μ [μ_N]	0.403(7)	0.402(15)	0.400(11)

culation of the same quantities with different volumes is necessary, however, since this approach is currently beyond our computational ability, we investigate the significance of such effects by changing the m_π while keeping the spatial extent, L, constant.

We compare the results of spin-1/2 Ω_c and Ω_{cc} baryons obtained on ensemble N5 to the ones calculated on N4 and N3 for which $m_\pi L = 4.3$ and $m_\pi L = 6$, respectively. We compile the numerical results into Table 5.20 and plot them in Fig. 5.24 for a clear comparison. Agreement within 1σ error bars suggests that the finite size effects are not severe for the charmed observables most probably due to small correlation lengths (larger masses) of charmed baryons.

5.5.3 *Fitting Procedures*

In the spin-1/2 sector, we extract the electric charge radii via dipole fits to form factors but change our approach for the spin-3/2 case. Rather than computing the form factors up to $Q^2 \sim 1.6$ GeV, we use only the $[Q^2 = 0, Q^2 = 1]$ momentum set as outlined in Sect. 5.4.1.2. Underlying assumption is that the form factors are well described by a dipole form hence Eq. (5.22) provides a reliable description. We check, however, if this assumptions holds by comparing the charge radii of the spin-1/2 Ω_c baryon extracted via a dipole-fit procedure and by Eq. (5.22). Results are illustrated in Table 5.14. Based on the agreement between the results, we conclude that both approaches agree and no systematic error arises due to the change of analysis.

Different courses of action in combining the data sets might lead to varying results. A good example is how one combines the data sets of individual quark sectors to estimate the properties of the baryon. In one approach, say approach I, we can extract the observables by first combining the data sets of the quark sectors which is followed by a fit to that combined data set. In approach II, on the other hand, same properties can be evaluated by performing fits to the data sets of individual quark sectors beforehand and then by combining those values via Eq. (5.18). In principle, both approaches should give the same value, however, statistical fluctuations may cause deviations. Correlations amongst the data are usually considered problematic but in approach I, it would help to cancel the similar statistical fluctuations of different observables, providing a better estimated value, especially if the value is close to zero like in the case of $\langle r_E^2 \rangle_{\Omega_{cc}}$. When possible, we always follow approach I, e.g. spin-1/2 values quoted in Tables 5.3, 5.11 and 5.17. An exception is the electric charge radius of zero electric-charge baryons since approach I naturally leads to an ill-defined form factor in a sense that it can not be described by a dipole form. In such cases (e.g. $\langle r_E^2 \rangle_{\Omega_c^0}$) we utilise approach II.

Agreement between the two approaches is visible in our numerical results. For example the magnetic moment of charged Ω_{cc} is estimated as $\mu_{\Omega_{cc}} = 0.403(7)\ \mu_N$ in approach I, while approach II results in $\mu_{\Omega_{cc}} = 0.422(10)\ \mu_N$, and the neutral Ω_c is $\mu_{\Omega_c} = -0.688(31)\ \mu_N$ in approach I, while approach II gives $\mu_{\Omega_c} = -0.714(35)$ μ_N, suggesting minimal systematic errors.

5.5.4 *Quark Action Related Uncertainties*

Quantities computed by employing the Clover action has $\mathcal{O}(m_q a)$ discretisation errors which should be controlled when one calculates observables related to heavy quarks, $m_q \gg \Lambda_{QCD}$. In this work, we have focused on charmed baryons composed of one or more valence charm quarks thus we check for the systematic effects due to Clover action. As discussed in Sect. 4.2.2, we set the improvement coefficient to its tad-pole improved value, $c_{SW} = 1/u_0^3$, following the procedure outlined in Ref. [35].

Note that on a set of ensembles all having the same lattice spacing, we have only indirect probes to discuss discretisation errors. A dedicated analysis with different lattice spacings is needed to quantify to which extent our results are prone to such errors. We simply argue that the systematic uncertainty on the observables is negligible compared to the current statistical precision.

5.5.4.1 Baryon Masses

We have presented and discussed our baryon masses in Sect. 5.1. Compared to the masses reported by other groups (see Table 5.1 and Figs. 5.3 and 5.4), we either see an agreement within error bars or—for charmed baryons—at most ~4% discrepancy. Note that other determinations employ relativistic heavy-quark actions and either make calculations on the physical point or perform chiral perturbation theory mandated fit forms to extrapolate their results to the physical point. A visible trend in Fig. 5.4 shows the degree of agreement improves as the number of valence charm quarks increase. The agreement in case of the Ω_{ccc} baryon is a good indication that the systematic errors associated with the c quark is at a minimum.

The small discrepancy might be attributed to our choosing $\kappa_{\text{val}}^s = \kappa_{\text{sea}}^s = 0.13640$, which, compared to its experimental value, leads to an overestimation of the mass of Ω (sss) baryon approximately by 100 MeV on ensemble N5 [6]. Retuning of κ_s so as to obtain the physical K mass would be desirable for precision calculations, however, choosing $\kappa^s_{\text{val}} \neq \kappa^s_{\text{sea}}$ would lead to partial-quenching effects and furthermore a retuning would have a minimal effect on the conclusions of this work.

5.5.4.2 Electromagnetic Observables

Based on the open-charmed meson form factor calculations [9], we expect the systematic effects to be much smaller for quantities that are less sensitive to the charm-quark mass. We may investigate further by mildly changing the κ_c from its determined value and check how the observables are effected. As long as we confirm that deviations do not affect the observables, it is safe to assume that the systematic errors are under control.

We first conduct a check in the spin-1/2 sector by comparing our results to a prior work on the Ξ_{cc} baryon with $\kappa_c = 0.1224$ and present recalculation with $\kappa_c = 0.1232$, both corresponding to approximately 100 MeV deviation in $m_{\Xi_{cc}}$. Results in Table 5.5 show virtually no change in the extrapolated values.

We have performed a similar check in the spin-3/2 sector also. We have repeated our simulations with $\kappa_c = 0.1256$, which leads to a decrease in the spin-3/2 charmed baryon masses by approximately 100 MeV from those given in Table 5.1. $E0$ and $M1$ form factor values are affected less than 1% and ~3% by such a change. Electric charge radii and magnetic moments are, in turn, effected by less than 3%. We find, for instance, $\langle r_E^2, \Omega_{ccc}\rangle_{\kappa_c=0.1246} = 0.170(9)$ fm^2 as compared to $\langle r_E^2, \Omega_{ccc}\rangle_{\kappa_c=0.1256} = 0.175(10)$ fm^2 and $\mu_{\Omega_c^*}^{\kappa_c=0.1246} = -0.696(50)\,\mu_N$ as compared to $\mu_{\Omega_c^*}^{\kappa_c=0.1256} = -0.712(50)\,\mu_N$ for 43 measurements on ensemble N5.

References

1. S. Prelovsek, Hadron Spectroscopy. PoS, **LATTICE2014**, 015 (2014)
2. Y. Namekawa, S. Aoki, K.-I. Ishikawa, N. Ishizuka, K. Kanaya, Y. Kuramashi, M. Okawa, Y. Taniguchi, A. Ukawa, N. Ukita, T. Yoshié, Charmed baryons at the physical point in 2+1 flavor lattice qcd. Phys. Rev. D **87**, 094512 (2013). https://doi.org/10.1103/PhysRevD.87.094512
3. C. Alexandrou, V. Drach, K. Jansen, C. Kallidonis, G. Koutsou, Baryon spectrum with $N_f = 2+1+1$ twisted mass fermions. Phys. Rev. D **90**(7), 074501 (2014). https://doi.org/10.1103/PhysRevD.90.074501
4. R.A. Briceno, H.-W. Lin, D.R. Bolton, Charmed-Baryon Spectroscopy from Lattice QCD with $N_f = 2+1+1$ Flavors. Phys.Rev. D **86**, 094504 (2012). https://doi.org/10.1103/PhysRevD.86.094504
5. Z.S. Brown, W. Detmold, S. Meinel, K. Orginos, Charmed bottom baryon spectroscopy from lattice QCD. Phys. Rev. D **90**(9), 094507 (2014). https://doi.org/10.1103/PhysRevD.90.094507
6. S. Aoki, K.-I. Ishikawa, N. Ishizuka, T. Izubuchi, D. Kadoh, K. Kanaya, Y. Kuramashi, Y. Namekawa, M. Okawa, Y. Taniguchi, A. Ukawa, N. Ukita, T. Yoshie, 2+1 flavor lattice QCD toward the physical point. Phys. Rev. D **79**, 034503 (2009). https://doi.org/10.1103/PhysRevD.79.034503
7. B.C. Tiburzi, A. Walker-Loud, Hyperons in two flavor chiral perturbation theory. Phys. Lett. B **669**, 246–253 (2008). https://doi.org/10.1016/j.physletb.2008.09.054

8. B.C. Tiburzi, A. Walker-Loud, Strong isospin breaking in the nucleon and Delta masses. Nucl. Phys. A **764**, 274–302 (2006). https://doi.org/10.1016/j.nuclphysa.2005.08.013
9. K.U. Can, G. Erkol, M. Oka, A. Ozpineci, T.T. Takahashi, Vector and axial-vector couplings of D and D* mesons in 2+1 flavor Lattice QCD. Phys. Lett. B **719**, 103–109 (2013a). https://doi.org/10.1016/j.physletb.2012.12.050
10. T. Bhattacharya, S.D. Cohen, R. Gupta, A. Joseph, H.-W. Lin, B. Yoon, Nucleon charges and electromagnetic form factors from 2+1+1-flavor lattice QCD. Phys. Rev. D **89**(9), 094502 (2014). https://doi.org/10.1103/PhysRevD.89.094502
11. L. Maiani, G. Martinelli, M.L. Paciello, B. Taglienti, Scalar densities and baryon mass differences in lattice QCD with wilson fermions. Nucl. Phys. B **293**, 420 (1987). https://doi.org/10.1016/0550-3213(87)90078-2
12. D.B. Leinweber, T. Draper, R.M. Woloshyn, Decuplet baryon structure from lattice qcd. Phys. Rev. D **46**, 3067–3085 (1992). https://doi.org/10.1103/PhysRevD.46.3067
13. D.B. Leinweber, T. Draper, R.M. Woloshyn, Baryon octet to decuplet electromagnetic transitions. Phys.Rev. D **48**, 2230–2249 (1993). https://doi.org/10.1103/PhysRevD.48.2230
14. S. Boinepalli, D.B. Leinweber, A.G. Williams, J.M. Zanotti, J.B. Zhang, Precision electromagnetic structure of octet baryons in the chiral regime. Phys. Rev. D **74**, 093005 (2006). https://doi.org/10.1103/PhysRevD.74.093005
15. S. Boinepalli, D.B. Leinweber, P.J. Moran, A.G. Williams, J.M. Zanotti, J.B. Zhang, Electromagnetic structure of decuplet baryons towards the chiral regime. Phys. Rev. D **80**, 054505 (2009). https://doi.org/10.1103/PhysRevD.80.054505
16. K.U. Can, G. Erkol, B. Isildak, M. Oka, T.T. Takahashi, Electromagnetic structure of charmed baryons in lattice qcd. J. High Energy Phys. **2014**(5), 125 (2014). ISSN 1029-8479. https://doi.org/10.1007/JHEP05(2014)125
17. C. Alexandrou, M. Brinet, J. Carbonell, M. Constantinou, P.A. Harraud, P. Guichon, K. Jansen, T. Korzec, M. Papinutto, Axial Nucleon form factors from lattice QCD. Phys. Rev. D **83**, 045010 (2011a). https://doi.org/10.1103/PhysRevD.83.045010
18. C. Alexandrou, M. Brinet, J. Carbonell, M. Constantinou, P.A. Harraud et al. Nucleon electromagnetic form factors in twisted mass lattice QCD. Phys.Rev. D **83**, 094502 (2011b). https://doi.org/10.1103/PhysRevD.83.094502
19. C. Alexandrou, T. Korzec, G. Koutsou, J.W. Negele, Y. Proestos, The electromagnetic form factors of the Ω^- in lattice QCD. Phys.Rev. D **82**, 034504 (2010). https://doi.org/10.1103/PhysRevD.82.034504
20. K.A. Olive, Particle Data Group, Review of particle physics. Chinese Phys. C **38**(9), 090001 (2014), http://stacks.iop.org/1674-1137/38/i=9/a=090001
21. A. Yamamoto, H. Suganuma, Quark motional effects on the inter-quark potential in baryons. Phys.Rev. D. **77**, 014036 (2008). https://doi.org/10.1103/PhysRevD.77.014036
22. K.U. Can, G. Erkol, B. Isildak, M. Oka, T.T. Takahashi, Electromagnetic properties of doubly charmed baryons in lattice qcd. Phys. Lett. B **726**(4–5), 703–709 (2013b). ISSN 0370-2693. https://doi.org/10.1016/j.physletb.2013.09.024, http://www.sciencedirect.com/science/article/pii/S0370269313007417
23. B. Julia-Diaz, D.O. Riska, Baryon magnetic moments in relativistic quark models. Nucl. Phys. A **739**, 69–88 (2004). https://doi.org/10.1016/j.nuclphysa.2004.03.078
24. A. Faessler, Th. Gutsche, M.A. Ivanov, J.G. Korner, V.E. Lyubovitskij et al, Magnetic moments of heavy baryons in the relativistic three-quark model. Phys. Rev. D **73**, 094013 (2006). https://doi.org/10.1103/PhysRevD.73.094013
25. C. Albertus, E. Hernandez, J. Nieves, J.M. Verde-Velasco, Static properties and semileptonic decays of doubly heavy baryons in a nonrelativistic quark model. Eur. Phys. J. A **32**, 183–199 (2007). https://doi.org/10.1140/epja/i2007-10364-y
26. A. Bernotas, V. Simonis, Magnetic moments of heavy baryons in the bag model reexamined (2012), arXiv:1209.2900
27. N. Sharma, H. Dahiya, P.K. Chatley, M. Gupta, Spin $1/2^+$, spin $3/2^+$ and transition magnetic moments of low lying and charmed baryons. Phys. Rev. D **81**, 073001 (2010). https://doi.org/10.1103/PhysRevD.81.073001

28. N. Barik, M. Das, Magnetic moments of confined quarks and baryons in an independent-quark model based on Dirac equation with power-law potential. Phys. Rev. D **28**, 2823–2829 (1983). https://doi.org/10.1103/PhysRevD.28.2823
29. S. Kumar, R. Dhir, R.C. Verma, Magnetic moments of charm baryons using effective mass and screened charge of quarks. J. Phys. G **31**, 141–147 (2005). https://doi.org/10.1088/0954-3899/31/2/006
30. B. Patel, A.K. Rai, P.C Vinodkumar, Masses and magnetic moments of heavy flavour baryons in hyper central model. J. Phys. G **35**, 065001 (2008). https://doi.org/10.1088/1742-6596/110/12/122010
31. S.-L. Zhu, W-Y.P. Hwang, Z.-S. Yang, Σ_c and Λ_c magnetic moments from QCD spectral sum rules. Phys. Rev. D **56**, 7273–7275 (1997). https://doi.org/10.1103/PhysRevD.56.7273
32. K.U. Can, G. Erkol, M. Oka, T.T. Takahashi, Look inside charmed-strange baryons from lattice QCD. Phys. Rev. D **92**(11), 114515 (2015). https://doi.org/10.1103/PhysRevD.92.114515
33. C. Aubin, K. Orginos, V. Pascalutsa, M. Vanderhaeghen, Lattice calculation of the magnetic moments of Δ and Ω^- baryons with dynamical clover fermions. Phys. Rev. D **79**, 051502 (2009). https://doi.org/10.1103/PhysRevD.79.051502
34. M. Luscher, Selected topics in lattice field theory. Conf. Proc. **C880628**, 451–528 (1988)
35. D. Mohler, R.M. Woloshyn, D and D_s meson spectroscopy. Phys. Rev. D **84**, 054505 (2011). https://doi.org/10.1103/PhysRevD.84.054505

Part III
Concluding Remarks

Chapter 6
Summary, Conclusions and Prospects

Abstract We summarize the main findings of this work in this final chapter and discuss the prospective future directions along with our on-going efforts.

Keywords Charmed baryons · Charge radii and magnetic moments
Baryon structure

6.1 Summary and Conclusions

Electromagnetic form factors is one of the pieces that play an important role in describing the internal dynamics of hadrons. They reveal valuable information about the size and the shape of the hadrons. Determining these form factors is an important step in our understanding of the hadron properties in terms of quark-gluon degrees of freedom. There have been enormous efforts to determine the electromagnetic form factors of light hadrons. The theoretical challenge is to understand these quantities from QCD. One intriguing question is how the structure of the hadrons gets modified in the heavy-quark regime, like in the case of charm hadrons. While there exist experimental results for the light baryons revealing their spectrum and electromagnetic properties, only the spectrum of the charmed baryons are accessible by experiments for the time being. Future experimental efforts at facilities like e.g. J-PARC, SuperKEKB, BES-III etc., are expected to provide a wealth of information, which calls for a better understanding of the heavy-sector dynamics from theoretical grounds. Combined with the available information on the light sector, insights on the heavy-flavor hadrons would reveal differences in the quark-gluon dynamics of heavy flavors.

In the framework of lattice QCD—the only known method that starts directly from QCD Lagrangian—the electromagnetic form factors of light hadrons have been extensively studied. Lattice computations have now reached an advanced level so that the simulations with (almost) physical light quarks are possible. Main challenges for the lattice QCD form factor calculations have been the pseudoscalar/vector-meson

K. U. Can, *Electromagnetic Form Factors of Charmed Baryons in Lattice QCD*,
Springer Theses, https://doi.org/10.1007/978-981-10-8995-4_6

states and the nucleon while the octet and decuplet baryon structure had less attention. Lattice literature on charmed baryon electromagnetic form factors have been non-existent. Along this perspective, we have built upon our experience on open-charmed meson sector to extend our calculations to charmed baryons. In this thesis, we have presented the combined results of our works on baryons that contain at least one charm quark.

After reviewing the QCD and relation of its elementary degrees of freedom to the hadrons in Chap. 2, we have covered the theoretical formalism to calculate the electromagnetic form factors in Chap. 4 and given the details of our lattice setup. In a nutshell, we have calculated the electromagnetic form factors and related static observables of the triply-strange $\Omega\,(\frac{3}{2}^+)$ baryon and the charmed $\Sigma_c\,(\frac{1}{2}^+)$, $\Omega_c\,(\frac{1}{2}^+)$, $\Xi_{cc}\,(\frac{1}{2}^+)$, $\Omega_{cc}\,(\frac{1}{2}^+)$, $\Omega_c^*\,(\frac{3}{2}^+)$, $\Omega_{cc}^*\,(\frac{3}{2}^+)$, $\Omega_{ccc}\,(\frac{3}{2}^+)$ baryons in the lattice QCD framework. We have utilised five sets of gauge ensembles incorporating the dynamical effects of u/d and s quarks with varying light-quark masses down to almost their physical values. Strange and charm quark masses are fixed to their respective physical values. We have run our simulations on lattices of $(32a)^3 \times 64a$ in size with a fine enough lattice spacing of $a = 0.0907(13)$ fm capable of resolving the inner structure of charmed baryons and corresponding to a spacious enough volume of approximately $(2.9\,\text{fm})^3 \times 5.9\,\text{fm}$ to accommodate them while minimising the finite-size effects to a negligible extent.

Masses, electromagnetic form factors and the related observables of the baryons have been presented in Chap. 5. We have extracted the masses of the baryons by analysing the two-point lattice correlation functions. Masses of the baryons that contain light quarks are studied on ensembles with different light-quark masses and baryon masses are estimated by extrapolations to the physical light-quark mass. Baryon masses obtained on the ensemble with almost physical light-quark mass are observed to be almost identical to the extrapolated masses. Our masses lie approximately 100 MeV above the experimentally measured masses of $\Omega_c\,(\frac{1}{2}^+)$, $\Omega\,(\frac{3}{2}^+)$ and $\Omega_c^*\,(\frac{3}{2}^+)$ baryons while the agreement between our and other lattice groups' determinations improve, as evident in Fig. 5.4, as the number of valence charm quarks increase. Small discrepancies most likely arise from the simple extrapolation forms that we have employed to estimate the masses on the physical light-quark mass point in combination with the questionable strange quark-mass tuning of PACS-CS. A HBχPT inspired extrapolation function for instance reproduces the mass of the Σ_c baryon in good agreement with its experimental value indicating that the extracted charmed-light baryon masses have minimal systematic errors. Nevertheless, the significance of the masses of the baryons in form factor calculations lies in the kinematical terms which have minimal effect on observables. We have indeed studied the possible effects of the mass discrepancies on the form factors and our tests indicate that the final electromagnetic observables are affected by less than 2%.

Electromagnetic form factors are related to the static properties, such as electric charge radii, magnetisation densities and the magnetic and higher order moments of the baryons, each providing an aspect of the internal structure. In general our results indicate that the charmed baryons are compact—the magnitude of their observables is decreased—compared to their light-sector counterparts. We have extracted the contributions of the individual quarks that build-up the baryon to the electromagnetic observables and identified the reduced contribution of the charm quark as the main source of the decrease. Electric charge radii results indicate that the distribution of the light quark is larger than the charm quark indicating that the charm quark acts as a hard core and shifts the center of mass of the system towards itself. Negative charge radii values obtained for $\Omega_c^{(*)}$ are indicative of a positively charged charm core covered with a negatively charged strange quark distribution. Electric properties are mildly sensitive to the flavor of the quark and the spin composition of the baryon which can be inferred from the similarity of the charge radii of Ξ_{cc}^{++} $(\frac{1}{2}^+)$, Ω_{cc}^+ $(\frac{1}{2}^+)$ and $\Omega_{cc}^{*\,+}$ $(\frac{3}{2}^+)$ baryons. Magnetisation density follows a similar trend as well but its spread is larger compared to the electrical distribution. We could access the electric-quadrupole moments of spin-3/2 baryons only in our setup and were able to isolate a clear signal for the $\Omega_{cc}^{*\,+}$ $(\frac{3}{2}^+)$ and Ω_{ccc}^{++} $(\frac{3}{2}^+)$ baryons. Their negative values are indicative of an oblate distribution of the electric charge.

Magnetic moments are found to be smaller compared to that of their light counterparts due to the charm quarks as well. Individual quark sector analyses reveal that the doubly represented quarks have a significant effect on the magnetic moment of the baryon. Alignment of the quark spins with respect to each other can be deduced from the sign of their magnetic moments and we observe that the quark sectors are anti-aligned in spin-1/2 but aligned in spin-3/2 baryons. Contribution of the charm quark is systematically smaller than that of light quarks, once again, but its magnitude changes in different spin configurations where it is enhanced in spin-3/2 baryons. Contribution of the light quarks is enhanced as well. We have also identified the effects of the environment on an individual quark by contrasting our results with that of the octet and decuplet baryons. Strange quark seems to be more sensitive to the changes to its accompanying quarks than the charm quark. In general, results are found to be consonant with the qualitative expectations of quark model and heavy-quark symmetry, although there are apparent quantitative differences.

Also in Chap. 5, we have identified and analysed several sources of systematic errors that might affect the final results. Excited state contamination has been deemed as under control. We have checked the finite-volume effects that may arise in ensemble N5 and by comparing the results of N5 to those obtained on N3 and N4, we have confirmed that finite-volume effects are under control as well. We have run dedicated tests to ensure that the discretisation errors associated with the charm quark is under control and concluded that the systematic effect to the form factors is in a few percent level—negligible with respect to the statistical errors. Possible inconsistencies that might arise due to the different statistical analysis procedures are shown to be of no concern.

6.2 Future Prospects

This work has provided the first systematic lattice QCD study of the electromagnetic form factors of charmed baryons as a part of our current wider program. It is natural to extend such calculations to study the electromagnetic transition form factors from which transition magnetic moments or (partial) decay widths of baryons can be extracted. Considering the rather limited experimental results on charmed baryon quantities, reliable first principle calculations have the potential to give accurate predictions. Comparing the individual quark sector contributions to the observables of baryons with different flavor structures and wave functions would provide further insight to the flavor dynamics and symmetries. In addition to the electromagnetic (vector-current) interaction, different currents probe different properties of hadrons. Form factors corresponding to axial-vector or pseudoscalar currents for instance, can be related to the spin structure or more interestingly to the pion interactions which is an important ingredient of effective models and relevant to the singly-charmed baryon phenomenology. Our current program progresses along these directions with the hope that it will help improve our understanding of the structure and interactions of heavy-flavored baryons from first principles.

Curriculum Vitae

Kadir Utku CAN
Cell (JP): +81-80-1988-2231
Cell (TR): +90-533-661-6212
e-mail: kadirutku.can@riken.jp
RIKEN Nishina Center
2-1 Hirosawa, Wako
Saitama 351-0198
Japan

Work

2017—present	**Post-doc**:	*RIKEN* Nishina Center Strangeness Nuclear Physics Laboratory

Education

2014–2017	**Ph.D**:	*Tokyo Institute of Technology* Graduate School of Science and Engineering Physics Department
2013–2014	**Research Student**:	*Tokyo Institute of Technology* Graduate School of Science and Engineering Physics Department
2012–2013	**Ph.D** (incomplete):	*Middle East Technical University* Graduate School of Natural and Applied Sciences Physics Department
2010–2012	**M.S**:	*Middle East Technical University* Graduate School of Natural and Applied Sciences Physics Department C.GPA: 3.21 / 4.00
2005–2010	**B.S**:	*Middle East Technical University* Arts & Sciences Faculty Physics Department C.GPA: 2.72 / 4.00

K. U. Can, *Electromagnetic Form Factors of Charmed Baryons in Lattice QCD*,
Springer Theses, https://doi.org/10.1007/978-981-10-8995-4

Languages

- **Turkish**: Native
- **English**: Fluent
- **Japanese**: Intermediate

Computer Skills

Languages:

- FORTRAN 77/95
- C/C++
- LaTeX
- CUDA
- Python

Software:

- PYTHIA 6/8
- CERN ROOT
- Mathematica
- CompHEP/CalcHEP
- CHROMA Library
- GNU Octave

Projects

Research Assistant:

- 2011–2013 : Non-perturbative approaches to hadron interactions in Quantum Chromodynamics (TUBITAK[1] Project No: 110T245)

Research Interests

- Lattice QCD
- Hadron Structure
- Heavy Hadrons

Peer-Reviewed Papers

1. H. Bahtiyar, K.U. Can, G. Erkol, M. Oka, T.T. Takahashi, $\Xi_c\gamma \to \Xi'_c$ *transition in lattice QCD*, **Phys.Lett. B772 (2017) 121–126**, https://doi.org/10.1016/j.physletb.2017.06.022
2. K.U. Can, G. Erkol, M. Oka, T.T. Takahashi, $\Lambda_c\Sigma_c\pi$ *coupling and* $\Sigma_c \to \Lambda_c\pi$ *decay in lattice QCD*, **Phys.Lett. B768 (2017) 309–316**, https://doi.org/10.1016/j.physletb.2017.03.006
3. K.U. Can, G. Erkol, M. Oka, T.T. Takahashi, *A look inside charmed-strange baryons from lattice QCD*, **Phys.Rev. D92 (2015) no.11, 114515**, https://doi.org/10.1103/PhysRevD.92.114515
4. H. Bahtiyar, K.U. Can, G. Erkol, M. Oka, $\Omega_c\gamma \to \Omega_c^*$ *transition in lattice QCD*, **Phys.Lett. B747 (2015) 281–286**, https://doi.org/10.1016/j.physletb.2015.06.006
5. K.U. Can, A. Kusno, E. V. Mastropas and J. M. Zanotti, *Hadron Structure on the Lattice*, **Lect. Notes Phys. 889 (2015) 69–105**, https://doi.org/10.1007/978-3-319-08022-2_3

[1]The Scientific and Technological Research Council of Turkey, http://www.tubitak.gov.tr/en.

6. K.U. Can, G. Erkol, B. Isildak, M. Oka, T.T. Takahashi, *Electromagnetic structure of charmed baryons in Lattice QCD*, **JHEP05 (2014) 125**, https://doi.org/10.1007/JHEP05(2014)125
7. K.U. Can, G. Erkol, B. Isildak, M. Oka, T.T. Takahashi, *Electromagnetic properties of doubly charmed baryons in Lattice QCD*, **Phys.Lett. B726 (2013) 703–709**, https://doi.org/10.1016/j.physletb.2013.09.024
8. K.U. Can, G. Erkol, M. Oka, A. Ozpineci, T.T. Takahashi, *Vector and axial-vector couplings of D and D* mesons in 2+1 flavor Lattice QCD*, **Phys.Lett. B719 (2013) 103–109**, https://doi.org/10.1016/j.physletb.2012.12.050

Scholarships/Awards

Awards:

- 2017 Springer Thesis Award 2017
- 2009–2010 Spring METU High Honor Roll
- 2009–2010 Fall METU Honor Roll
- 2008–2009 Spring METU Honor Roll

Scholarships:

- 2014–2017 Japanese Government (MEXT) Scholarship as a Ph.D Student
- 2013–2014 Japanese Government (MEXT) Scholarship as a Research Student

References
Available upon request

Parts of this thesis have been published in the following articles:

- K.U. Can, G. Erkol, M. Oka, T.T. Takahashi, Look inside charmed-strange baryons from lattice QCD, Physical Review D 92, 114515 (2015).
- K.U. Can, G. Erkol, B. Isildak, M. Oka, T.T. Takahashi, Electromagnetic structure of charmed baryons in Lattice QCD, Journal of High Energy Physics (JHEP) 05, 125 (2014).
- K.U. Can, G. Erkol, B. Isildak, M. Oka, T.T. Takahashi, Electromagnetic properties of doubly charmed baryons in Lattice QCD, Physics Letters B726 (2013) 703–709.